Production Checklist For Builders And Superintendents

Concrete • Framing • Finish • Final

John J. Haasl and Peter Kuchinsky II

Home Builder Press
National Association of Home Builders

Production Checklist for Builders and Superintendents
ISBN 0-86718-351-9

Copyright 1990 by the National Association of Home
Builders of the United States

Printed in the United States of America

Library of Congress Cataloging in Publication Data
Haasl, John J., 1948-
   Production checklist for builders and
superintendents / John J. Haasl and Peter Kuchinsky II.
   p. cm
   ISBN 0-86718-351-9 (paper) : $20.00 ($16.00
NAHB members)
   1. Building—Handbooks, manuals, etc. 2.
Building—Forms.
I. Kuchinsky, Peter, 1959- . II. Title.
TH151.H22 1990
692—dc20

                                        90-4909
                                        CIP

For further information, please contact:
   Home Builder Bookstore
   15th and M Streets, N.W.
   Washington, D.C. 20005
   (800)368-5242

9/90 Scott/McNaughton & Gunn 2K
6/91 McNaughton & Gunn 2.5K

Table of Contents

About the Authors

John J. Haasl and Peter Kuchinsky II, authors of the *Production Checklist for Builders and Superintendents*, have over thirty-five years of combined experience in the construction industry. Their diversified backgrounds and experiences include numerous field, project, and executive positions, from superintendent to vice president of construction, with major developers and homebuilders.

Through this experience, they have been responsible for completing millions of dollars in improvements and delivering thousands of housing units. Both authors are fully licensed general contractors in the state of California. The production checklist was born not only as a working tool for the authors, but as a teaching aid for superintendents who have come under their direction.

Acknowledgments

John Haasl and Peter Kuchinsky are grateful for the comments and contributions of Jerry Chiles, Vice President, Wolf Pacific Companies; Al Dary, Field Supervisor, Pacific Telephone; Ian Gardiner, Vice President, Broadmoor Homes; Carol Hale, Customer Service Manager, Daly Homes; Terri Meinhart, Construction Secretary, Daly Homes; Ed Reeves, Owner, R & R Construction; Norm Schmitt, President, Construction Consultants, LTD; Jerry Householder, Director of Graduate Studies in Construction Management, Virginia Polytechnic Institute and State University; David Vanderslice, Vice President of Construction, Centennial Homes; Mike Milliner, President, Milliner Construction; Bob Whitten, Wayne Homes; and Andrew Cox, President, Building Inspection Services, Inc.

Thanks go out to all the people at the following companies for their contributions: Broadmoor Homes, California Communities, Inc., Daly Homes of California, Meeker Development, Stein-Brief Construction Group, and Universal Development Company.

The authors would also like to thank Mary Cox and Curt Hane of Home Builder Press, whose efforts helped to make this book a reality.

This book was produced under the general direction of Kent Colton, NAHB Executive Vice President, in association with NAHB staff members James E. Johnson, Jr., Staff Vice President, Operations and Information Services; Adrienne Ash, Assistant Staff Vice President, Publishing Services; Rosanne O'Connor, Director of Publications; Curt Hane, Assistant Director of Publications; Mary Cox, Publications Editor; and David Rhodes, Art Director.

1

Introduction

Over the past twenty-five years, Americans have become extremely quality conscious. They study various consumer publications prior to making major purchases. New safety features and additional quality controls are required to manufacture toasters, clocks, television sets, even pairs of jeans. Yet, it is amazing how little time and attention are given to quality control programs in residential construction, especially during the critical concrete and framing phases of structural construction. With recent shifts in the legal system from the theory of "buyer beware" to "builder beware," those builders who fail to institute a comprehensive quality control program could face costly litigation in the future.

The purpose of this book is to provide the construction and development industry with a comprehensive production checklist and deliver to superintendents a method of checks and balances to complete their projects with the highest quality and cost-effectiveness possible.

This checklist, and the interrelated construction schedules, were developed to familiarize new project superintendents with—and remind experienced supervisors of—the day-to-day, step-by-step construction procedures necessary to establish standards of quality, confirm start dates, monitor progress, predict completion dates, meet a production schedule, and, most importantly, produce a *quality* finished product.

Use of the *Production Checklist for Builders and Superintendents* cannot guarantee the success of any project. However, if committed to and used as a working tool, it will increase the quality control of any project, which will contribute to the overall success of the project and company.

This list has taken over twenty-five years to assemble. It is an accumulation of positive information produced from trial and error. It is a conglomeration of our best piecemealed punch lists, and a composite of the "unwritten" jobsite knowledge that can only be obtained through experience.

This book is unique because it contains the most comprehensive lists covering all aspects of on-site construction supervision. While it does not pretend to cover *every* possible situation (to do so would require an encyclopedia), it provides those who want to learn a source of information. Its purpose is to help improve quality in the building industry and maybe make your job, as a builder or superintendent, just a little easier.

These checklists and schedules were not designed for the "run-the-project-from-the-trailer" type of superintendent. They are intended for the "walking-and-working" project superintendent—the type of supervisor that wants to be involved and keep control of the

5

daily affairs of a construction project. Quality control must start the very first day of production and be the last task performed upon completion of a project.

It is unfortunate that many project supervisors choose to abandon quality control responsibilities during the rough and mechanical phases of construction and rely solely upon the building inspector to police the structural integrity, completeness, and overall quality of the project. When quality control programs start during or after the finish phase of construction, most major unsolved problems are likewise saved for the final hour. This procrastination and negligent style of supervision will probably end in failure. This type of project will most likely experience cost overruns and delivery delays.

This checklist represents an effort to stress the importance of quality throughout *every* phase of construction—concrete, framing, mechanical, finish, and final. If proper quality control standards are initiated at the very onset of the project, fast and accurate work can be performed while still meeting a realistic production schedule. However, pride of workmanship and quality should never be compromised in order to meet *any* schedule. Construction delays caused by the lack of proper planning can be eliminated by following a well-defined method of procedures and checklists.

These checklists and schedules are designed for slab-on-grade, poured-in-place, in-line, and production housing tracts. Because every area of the country builds houses and determines the appropriate schedules to meet the regional and product requirements, the steps and allotments of time may vary. This step-by-step process, through superintendent involvement, can be applied, with minor regional adjustments, to any project in any part of the country.

Safety Checklist

General Safety

1) _____ Maintain a fully equipped first-aid kit on the site at all times.
2) _____ Hold a construction safety meeting every ten working days. Keep notes on file in the trailer.
3) _____ Post and maintain "No Trespassing" signs around the perimeter of the project. Make sure they are no more than fifty feet apart.
4) _____ Post "911" and emergency numbers by all telephones.
5) _____ Maintain a list of all hazardous materials used on the project.
6) _____ Document and report all accidents and injuries on appropriate accident report forms.
7) _____ Make sure hard hats, hearing protection, and necessary safety equipment are used by all workers.
8) _____ Keep the jobsite clean. Remove scrap materials and trash from the site regularly.
9) _____ Make sure all heavy equipment and job vehicles have back-up signal alarms.
10) _____ Make sure all excavation is properly protected by barricades and safety reflectors. Use flagmen for traffic control as needed. Make sure flagmen wear red or orange reflectorized garments.
11) _____ Make sure all delivered materials are stacked and secured. Do not obstruct driveways, roads, passageways, or exits.
12) _____ Make sure fire extinguishers and required fire equipment are certified and maintained.
13) _____ Review jobsite policies with all subcontractors prior to the start of work and at safety meetings:
    A) _____ Define working hours and days.
    B) _____ Explain site protection and traffic control policies.
    C) _____ Establish emergency procedures for accidents, fire, or injuries.
    D) _____ Clarify cleanup responsibilities.
    E) _____ Set standards for worker's behavior and attitudes.
    F) _____ Emphasize a no drug or alcohol policy.
    G) _____ Establish that no children or pets will be allowed on the jobsite. Set up an authorization system for jobsite visitors.

Note: Most insurance carriers will be more than happy to provide the necessary forms, procedures, formats, and the pertinent data required to conduct weekly "tailgate" safety meetings. If you take the time to ask, they will gladly provide you with everything needed to establish a complete safety program.

Tools and Equipment

1) _____ Inspect hand tools and power equipment regularly for defects or unsafe conditions.

2) _____ Make sure only trained personnel use power tools. Make sure proper eye, face, and safety equipment is used with power tools.

3) _____ Make sure all power tools and electrical cords are double-insulated or properly grounded.

4) _____ Make sure powder activated tools are used only by trained personnel and left unloaded.

5) _____ Make sure pneumatic tools, hoses, and attachments are securely fastened. Make sure their recommended pressure is not exceeded.

6) _____ Make sure ladders are adequate for their purpose, in good condition, have secure footing, and are protected against movement.

7) _____ Make sure temporary electrical power poles are approved and in good condition. Make sure power lines are properly protected and grounded with G.F.I. circuits. Make sure power lines are kept ten feet away from all buildings, ladders, and scaffolding. Make sure power lines crossing streets are at least twenty feet above the road to prevent damage from heavy equipment, cranes, and delivery trucks.

8) _____ Make sure extension cords are the three-wire type, enclosed in protective sheathing.

9) _____ Make sure only qualified personnel operate cranes, pettibones, and other heavy equipment. Make sure equipment is used for only its intended purpose. Make sure rated load capacities and operating speeds are not exceeded.

Scaffolding, Wall, and Floor Openings

1) _____ Make sure scaffolding is adequate for its intended purpose, and in good condition. Make sure footings are protected against movement. Make sure the frame is sound, rigid, and capable of supporting the intended load.

2) _____ Make sure planking does not extend less than six or more than twelve inches over the ends of frames. Make sure planking overlaps at least twelve inches.

3) _____ Make sure poles, legs, and frames are plumb, with adequate cross bracing to prevent swaying. Make sure scaffolding is secured to the building to prevent it from falling down or becoming displaced.

4) _____ Make sure guardrails and toeboards are provided on any scaffold, wall, or floor opening six feet or more above the floor or ground. Guardrails should be forty-two inches above the floor or planking.

5) _____ Make sure guardrail can withstand a minimum impact of two hundred pounds.

Concrete and Framing

1) _____ Make sure formwork, shoring, and framing are adequate to support all intended loads during the course of construction.

2) _____ Make sure protruding rebar or framing is bent, covered, or adequately protected.

3) _____ Make sure protruding nails are stripped or bent over at all times to prevent injury.

Flammable Materials and Gases

1) _____ Identify all agents that may cause injury through inhalation, ingestion, or skin and eye contact. Make sure proper safety precautions are followed. Educate all personnel on the hazards of using harmful agents.

2) _____ Check for adequate ventilation (natural or mechanical).

3) _____ Make sure face masks or respirators are used by workers who are exposed to excessive dust or harmful gases.

4) _____ Store flammable or combustible materials in approved containers. Post "No Smoking" and "No Open Flame" warning posters. Check OSHA and local regulations when storing over twenty-five gallons of flammable or combustible materials.

5) _____ Make sure proper fire precautions are used during welding and torch operations. Make sure only qualified personnel use welding and torch equipment. Make sure there is a fire extinguisher in the immediate area.

6) _____ Make sure any portable heaters have automatic shut-off devices in case of flame failure or knock-over.

Excavation

1) _____ Notify all public utilities at least forty-eight hours prior to the start of any underground trenching or excavation work. Make sure they identify and mark all existing pipes and utility lines.

2) _____ Make sure trench shoring is used to support walls and faces of excavations exceeding five feet. If workers are entering and working in trenches, make sure all loose material is kept back a minimum of two feet. Make sure workers use ladders to enter trenches or excavations deeper than four feet.

3

Concrete Production

Introduction

Rigorous quality control during the concrete operation is of the utmost importance. Not only are concrete labor and materials very costly, but concrete buries other contractors' materials. Once placed, concrete is unworkable, except with a jackhammer or bulldozer, and this type of a repair could easily be considered a major correction. If the foundation is accurate—true to line and grade—the other trades, such as masonry, framing, plumbing, and drywall, stand an excellent chance of following the standards of quality established by the concrete operation.

Likewise, if the concrete is of poor quality structurally and not true to line and grade, it is almost impossible for a builder or project superintendent to expect the other subtrades to deviate from the architectural plans and modify their operations to correct any errors and omissions committed during the concrete operation.

Prior to the start of the concrete operation, the project superintendent should meet, at the jobsite, with the various subcontractors and the local building inspector to discuss inspection procedures, new codes, local code upgrades, subcontractor responsibilities, schedules, safety, and work hours. The building inspector's favorite "gotchas" or problem areas that will require special attention and receive close inspection should be discussed at this meeting.

Precise layout is crucial to the entire concrete operation. Prior to starting, the builder or project superintendent must confirm that the plans being used are the "final-approved" set. They must coincide with all other improvement plans, such as grading, drainage, street improvements, utilities, retaining walls, landscape, final grading, and any other plans that may exist or could affect the project.

The builder or superintendent must then validate the building pad elevations; reestablish property lines; verify set-backs; confirm the proper positioning and facing of the building; consider retaining walls, drainage devices, and positive lot drainage; and finally, approve the correct locations and depths of footings. The layout must be checked and rechecked to avoid any possibility of error.

It is not uncommon for a foundation to be laid out and poured on the wrong lot, or placed too close to the property line. Occasionally, the front elevation is built facing the rear property line, or the building footprint is reversed. The wrong or unapproved plan can even be applied to the right lot. Sometimes the house just does not fit on the lot. Such mistakes cannot be justified. Not only are they costly, but also very embarrassing. These mistakes can be avoided by checking and rechecking the approved plans and then applying the information on the plans to the actual site conditions.

Several well spent hours reading the plans and specifications prior to the start of work and familiarizing yourself with all contracts and schedules can save valuable time and thousands of dollars once production has begun. The following concrete checklist, when put to use *daily* on the jobsite, will help the supervisor avoid some of the pitfalls associated with concrete operations.

As previously stated, the concrete operation establishes the pace and quality of those subtrades that follow. Time spent checking dimensions, layout, and details prior to pouring and placing concrete can save both valuable time and dollars throughout the course of the project.

Concrete Checklist

Protection of Concrete Work

1) _____ Before the start of the concrete operation, the superintendent and the concrete foreman should walk through all jobsite improvements and identify any existing damage, both in writing and by means of fluorescent spray paint. The improvements checked should include finish lot grading, drainage devices, walls, landscaping, curbs, gutters, sidewalks, driveway approaches, asphalt, and utilities. When the concrete phase of construction has been completed, another walk should be scheduled to identify any new damage caused by the concrete operation.
First walk for damage complete: Yes ☐   No ☐
Concrete subcontractor's initials: _____ Date _____
Final walk for damage complete: Yes ☐   No ☐
Concrete subcontractor's initials: _____ Date _____

2) _____ Prior to the start of the project, establish written records on the following: daily diary, phone log, "notice to proceed" letters, daily concrete pour sheets, inspection requests and inspection log, daily production folder, second tier subcontractor report, and subcontractor rating report.

3) _____ Check with the grading and public works inspectors to make sure that grading has been passed and is ready to trench. Next, walk the tract and make sure all debris, such as excess sewer and water pipe, is picked up and removed. Make sure damaged pads are put back into shape, and that any piles of excess sand used for backfilling are removed. Remove all large rocks or chunks of concrete. Provide access to the pads for the trencher.

4) _____ Study the plans; know what you are building. Make sure all blue tops, building corner stakes, step-down stakes, slopes, stairs, walls, and garage stakes are installed.

5) _____ Make sure pads are graded for the proper type of building.

6) _____ Check all building pad area drains to ensure that all drains are protected with sand bags or plywood. Clearly mark their locations.

7) ___ Install building signs with lot numbers and type of plan identifications. Identify each lot.

8) ___ Order chemical toilets.

9) ___ Establish the start date. Confirm procedures for purchase orders and subcontractor billings.

10) ___ Schedule the first inspection date so building permits can be obtained and plans approved.

11) ___ Order all reports and plans—soils report, grading plans, stairway and retaining wall plans, joint trench plans, landscape plans, architectural plans, and street improvement plans. Read everything in the filing cabinets. Date all correspondence prior to filing. Everything in the filing cabinet must have a date stamp!

12) ___ Send out "notice to proceed" letters to concrete, framing, plumbing, electrical, masonry, and temporary power subcontractors.

13) ___ Check the layout and trenching of building pads. Have all excess dirt removed.

14) ___ After forms are set, ask engineer to verify the correct orientation of building on pad and to confirm that it has the proper clearances.

15) ___ Check forms for right elevation, width, and depth of footings and piers. Check for the right set backs. Tape all measurements for length and width of building and all interior bearing and non-bearing footings and piers. After step-down footings, landings, and slope footings are formed, set up builder's level and double-check elevations before pouring. Make sure forms are well braced, tight, free of debris, and coated with approved compound for ease of removal. Check grading plans for deepened footings and raised stem walls.

16) ___ When soil pipe is set, make sure there is a minimum one-eighth-inch clearance to the foot fall. Obtain a set of isometric drawings from the plumber so pipe size and location can be double-checked. File a copy for future verification of problems, if any. Also, fill pipes with water to test for leaks.

17) ___ Check electrical, telephone, television, gas sleeve, and ground rod locations. Check steel to verify proper size and location. Make sure steel is properly hung and clean prior to county or city footing inspection.

18) ___ Order water meter jumpers, tail pieces, and washers from the water district. Check street improvement plans to verify size of services.

19) ___ When pouring footings, check for the following:

A) ___ Read all contracts thoroughly so you will know just what is going in and under slabs. Check for proper thickness of slabs, correct steel, and proper concrete mix to ensure that the concrete reaches the required P.S.I.

B) ___ Get a ticket for every load of concrete poured. Check for the proper type of concrete, cement sack mix, and water ratio. Make sure it is per contract and soils report. Fill out daily pour sheets. Take cylinder samples and slump tests, as required.

C) ___ Check anchor bolt, hold-down, post anchor, column base, and post base locations for proper sizes and heights.

D) ___ Make sure all depths are correct per plans and soils report. Make sure footing trenches and rebar are three inches away from the forms and dirt banks. Make sure trenches are free of debris and dirt clods.

E) ___ Check for eased edges at stem walls and step downs, as required.

F) ____ Check installation of special expansion joints at split level, step, or grade break areas.

G) ____ Make sure concrete is placed properly.

H) ____ Strip and stack forms neatly.

I) ____ Fill and patch all rock pockets as soon as possible after forms are removed.

20) ____ When grading for the slab area, make sure all footings, including interior, are exposed, and cleaned of all loose dirt. Keep visqueen off all footings. Check mill size of visqueen. Make sure outside footings are three inches away from forms. Check soils report to see if presaturation is necessary on any pads. Prior to visqueen and sand or gravel fill, the copper water lines should be installed. Check for the following:

A) ____ Confirm proper type and size of copper line: type "K" or "L."

B) ____ Make sure there are no splices under the slab.

C) ____ Make sure no copper is exposed to concrete and that all copper installed has approved plastic sleeves to prevent electrolysis.

D) ____ Recheck forms for correct elevations. Check all buildings for backfill with the proper sand or gravel at raised slab area. Check visqueen—repair any holes or rips. Next, sand or gravel fill over visqueen, if required. Check sand for proper coverage (one or two inches) and for proper moisture content. String line all slab areas for four-inch nominal slab. Always wet sand, especially if it is a hot day. Make sure interior trenches are wheel rolled and filled with good native soil or import sand, whichever is required. Make sure plumber's trenches have double-wire mesh (if required). Make sure wire mesh is per contract and soils report. Check electrical conduit and floor boxes. Make sure all plumbing and electrical is tied down to ensure that it will not move or float during the pour. Schedule a pre-slab inspection.

21) ____ When pouring house and garage slabs, check the following:

A) ____ Check the concrete load tickets for the proper type of concrete, cement sack mix, and water ratio.

B) ____ Make sure the concrete is placed properly and not poured too wet.

C) ____ Take cylinders and required slump tests.

D) ____ Make sure mesh is pulled up when pouring and that chairs (if required) are installed. Make sure the "hook-man" does not stand or walk on mesh after it has been centered in the slab. If mesh is hooked and centered, do not allow workmen to stand or walk on it.

E) ____ Make sure tops of footings are clean. They should be swept or hosed off prior to the pour.

F) ____ Check for screed pins or grade stakes at proper spacing.

G) ____ Make sure all screed boards are straight.

H) ____ Check for even screed—no highs or lows.

I) ____ Make sure slab has been evenly screeded and tamped with three-eighths-inch mortar at the surface.

J) ____ Check for control joints as required (tooled, pre-formed, saw cut).

K) ____ Check for proper spacing of anchor bolts. Make sure there are no anchor bolts in any door openings.

L) _____ Make sure imbedded hardware is located per plan.

M) _____ Make sure bull nose is correct per plan.

N) _____ Spray with approved curing compound.

Stem Walls and Poured-In-Place Basements

1) _____ Determine at the prejob meeting if excavation is needed to accommodate building plans.

2) _____ Confirm who is responsible for excavation. Review contracts.

    A) _____ Confirm type of excavating equipment to excavate. Make sure the equipment chosen is the most efficient and productive.

    B) _____ Determine who is responsible for the export of dirt and excess material:
Owner _____ Subcontractor _____ .

    C) _____ Have the civil engineer determine the quantity of excess dirt prior to the start of export operations.

    D) _____ Decide which type of trucks will be needed for the haul.

    E) _____ Confirm the location of the dump site or stockpile.

3) _____ Before the dirt is exported:

    A) _____ Determine how many trucks will be needed.

    B) _____ Estimate how much loading time will be needed; include stand-by time waiting to load.

    C) _____ Check cycle time (round trip—site to dump site and back).

    D) _____ Keep a daily log and load count. Document and confirm load counts at the end of each day. Make sure the counts from the jobsite, trucking operation, and export site agree.

4) _____ Establish pad height and correct excavation. Have engineers stake and confirm.

5) _____ Lay out and trench for footings and piers.

6) _____ Set forms and have soils engineer inspect bottom of footing.

7) _____ Set steel. Install form ties. Brace outside of forms to prevent blowouts during the pour.

8) _____ Install conduits, sleeves, or head-outs required for future services, water line, soils pipe, beam pockets, or access.

9) _____ Lay out and install all hardware. Confirm locations with plans and concrete and framing contractors.

10) _____ Inspect steel and forms. Obtain permission to pour.

11) _____ Schedule concrete and double-check all layouts and dimensions.

12) _____ When pouring concrete, check for the following:

    A) _____ Read all contracts thoroughly. Check for correct height and thickness of walls, correct steel, and proper concrete mix to ensure that the concrete reaches the required P.S.I.

    B) _____ Get a ticket for every load of concrete poured. Check for the proper type of concrete, cement sack mix, and water ratio. Make sure it is correct according to the plans and soils report. Fill out a daily pour sheet. Take cylinder samples and slump test, as required.

C) _____ Check anchor bolt, hold-down, post anchor, column base, and post base locations for right sizes and heights.

D) _____ Make sure all depths are per plans and soils report. Make sure footing trenches and rebar are three inches away from the forms and dirt banks, and trenches are free of debris and dirt clods.

E) _____ Check for eased edges at stem walls and step downs, as required.

F) _____ Check installation of special expansion joints at split level, step, or grade break areas.

G) _____ Make sure concrete is placed properly.

H) _____ Make sure concrete stem walls are finished to a flat surface.

I) _____ Strip and stack forms neatly.

J) _____ Fill and patch all rock pockets as soon as possible after forms are removed.

13) _____ When grading for the slab area, make sure all footings, including interior, are exposed and cleaned of all loose dirt. Keep visqueen off all footings. Make sure outside footings are three inches away from forms. Check soils report to see if presaturation is necessary on any pads. Prior to visqueen and sand or gravel fill, the copper water lines should be installed. Check for the following:

A) _____ Confirm type and size of copper line: type "K" or "L."

B) _____ Make sure there are no joints under the slab.

C) _____ Make sure no copper is exposed to concrete and that all copper installed has approved plastic sleeves to prevent electrolysis.

D) _____ Recheck forms for correct elevations. Check all buildings for backfill with the proper sand or gravel at raised slab areas. Repair any holes or rips in visqueen. Next, sand or gravel fill over visqueen, if required. Check sand for proper coverage (one or two inches) and for proper moisture content. String all slab areas for four-inch nominal slab. Always wet sand, especially if it is a hot day. Make sure interior trenches are wheel rolled and filled with approved native soil or import sand, whichever is required. Make sure plumber's trenches have double-wire mesh (if required). Make sure wire mesh is per contract and soils report. Check electrical conduit and floor boxes. Make sure all plumbing and electrical is tied down to ensure that it does not move or float during the pour. Schedule a pre-slab inspection.

14) _____ When pouring house and garage slabs, check the following:

A) _____ Check the concrete load ticket for the proper type of concrete, cement sack mix, and water ratio.

B) _____ Make sure concrete is not poured too wet and is placed properly.

C) _____ Take cylinders and required slump tests.

D) _____ Make sure mesh is pulled up when pouring and chairs (if required) are installed. Make sure the "hook-man" does not stand or walk on mesh after it has been centered on the slab. If mesh is hooked and centered, do not allow workmen to stand or walk on it.

E) _____ Make sure the tops of footings are clean. They should be swept or hosed off prior to the pour.

F) _____ Check for proper spacing of screed pins or grade stakes.

G) _____ Make sure all screed boards are straight.

H) _____ Check for even screed—no highs or lows.

I) _____ Make sure slab is tamped with three-eighths-inch mortar at the surface.

J) _____ Check for control joints as required (tooled, pre-formed, and saw cut).

K) _____ Check for proper spacing of anchor bolts.

L) _____ Make sure imbedded hardware is located according to plan.

M) _____ Make sure bull nose is per plan.

N) _____ Spray with approved curing compound.

Stem Wall and Basement Backfill

1) _____ After forms are stripped and concrete has been cured (for a minimum of seventy-two hours) or concrete blocks have been laid, start backfill operation:

A) _____ Dig back from the foundation and expose twelve inches of footing.

B) _____ Remove all loose soil and material.

C) _____ Apply approved waterproofing membrane or system. Cover all areas completely, including exposed footing. Extend waterproofing six inches above the finish grade at the building line.

D) _____ Allow waterproofing to dry for twenty-four hours before backfilling.

E) _____ Install french drain system and crushed rock as required per plans.

F) _____ Start soil backfill operation. Be careful not to damage waterproofing. Compact soil as you backfill. Soil lifts should not exceed two-foot depths. Confirm compaction and backfill with soils engineer; compaction and backfill must be tested and certified in writing by soils engineer.

2) _____ Upon completion of backfill, cut and reestablish drainage away from stem walls and building.

Garages

1) _____ Check grading plans for special slab elevations.

A) _____ Check felt at edges.

B) _____ Check for three-inch slope.

C) _____ Clean exposed stem walls.

D) _____ Quarter slab with saw cut in twenty-four hours.

E) _____ Remove forms and debris.

F) _____ Complete steps and raised areas.

G) _____ Make sure nuts and washers are delivered in bulk and given to the framer.

H) _____ Use a ten-foot straight edge on slabs to check for highs and lows. Chip, grind, or fill accordingly.

I) _____ Water test slab for low areas (duck ponds). *Note*: Water aids in concrete curing process.

2) ____ When utilities are being installed, make sure you have gas stub-outs marked on footings so gas company will know where to run their line. *Note*: Prior to concrete, make sure gas, electric, and television sleeves are in place (usually in garages) at utility boxes or panels. Confirm locations of utilities prior to concrete pour.

3) ____ When house sewer laterals are installed and backfilled, schedule water services. Make sure proper type of pipe is used. Have it inspected by the municipality and backfilled. *Note*: All backfilled ditches should be at a minimum of 90 percent compaction. Schedule soils engineer to test and certify.

4) ____ After rough grade, check each pad for adequate drainage away from the building.

5) ____ When chipping for plaster screed, check the following:

A) ____ After framer raises first floor walls, walk perimeters of all buildings and garages and mark, with a felt pen on a tract map, areas that need chipping or dry pack. Make copies of highlighted tract maps. File one copy in the concrete contract folder and give other copies to the concrete foreman with a projected date of completion. *Note*: Fluorescent spray paint is excellent for identifying problem areas.

B) ____ Check-off the reason these areas need to be chipped:

1) ____ Concrete measured, laid out, or formed wrong.

2) ____ Frame measured or laid out wrong.

3) ____ Wrong plans; errors in dimensions; preliminary, outdated, or unapproved plans.

4) ____ Foundation plan dimensions used instead of floor plan dimensions.

5) ____ No confirmation of dimensions by subcontractor or superintendent prior to pouring concrete.

C) ____ Correct any problems as soon as possible.

Rough Trades Production

Introduction

Framing is the locomotive that pulls all other trades through the rough phases of construction. It establishes the character, style, and flow of the structure. The future homeowner *will* notice the results of rough carpentry. If walls, rafters, or joists are bowed, corners are not squared, or ceilings not leveled, numerous defects can occur. These defects will persist throughout the entire finish operation and remain after a home-owner occupies the structure. They become major difficulties that sooner or later *must* be addressed.

The unproclaimed prerequisite for project superintendents has traditionally been past carpentry experience. Many project superintendents were formerly rough carpenters, and this knowledge and experience can be helpful. The rough trades checklist can serve as an inspection tool for the experienced supervisor or a simple step-by-step guide for those lacking a rough carpentry background.

Too many superintendents have established themselves as jobsite baby-sitter, performing head counts and watching subtrades come and go. These superintendents fail to realize that they are also an inspector for the builder. They are directly responsible for inspecting and accepting all labor and materials that go into the building structure. By failing to perform quality control checks and demand corrections by the subtrades, the superintendent is the individual who "builds in" the defects seen by the homebuyer in the finished product.

This checklist will enable the superintendent not only to observe and monitor the quality control—but also to actively participate in the rough frame and mechanical phases of the project. When faithfully used, this checklist will help eliminate many of the problems that can later beleaguer and finally overwhelm the production homebuilder. Through repeated use, these lists should enable you to rapidly and clearly identify potential problem areas.

Rough Trades Checklist

1) ____ Order television cable and cans for electrical subcontractor from the cable television company.

2) ____ Contact all rough and mechanical subtrades. Confirm all framing rough openings and backing requirements in writing. File with contracts and provide copies to the framer. *Note*: Scaled-down floor plans or sales floor plans with handwritten, highlighted dimensions are an excellent way of controlling consistency. These will also eliminate confusion concerning locations of rough openings, backing, and cutouts for subcontractors.

Framing

1) ____ As soon as framer snaps lines, but before the mudsill or top plate is installed, check for the following:

   A) ____ Check all measurements (by taping); make sure all exterior walls are sitting on the slab. If they are not, determine what the problem is and correct it. *Note*: Is the concrete off? Did the layout man pull the wrong measurements? Do the framing measurements on the floor plan match the foundation plan?

   B) ____ Check anchor bolt and hold-down layouts. Break off any anchor bolts in doorways. If any hold-downs or post anchors are not working, now is the time to check and correct the measurements.

   C) ____ As soon as the layout man has detailed (all mudsills, top plates, headers, and cripples are cut and laying in place, ready for the framers), double-check all windows and door openings for correct sizes and locations.

2) ____ When raising walls, make sure there are two 16-D nails top and bottom on every stud, approximately three-quarters of an inch in from the ends. Check to see if mastic or roll caulking is required under mudsills. Make sure all let-in braces are installed where called for and properly nailed. Make sure posts, king studs, top plates, and trimmers are not cut too short. Make sure all upper and lower cripples for windows and doors are the proper length. Make sure the kitchen window height is correct for cabinet installation. Make sure the correct hold-downs are installed.

3) ____ Check plumb and line with an eight-foot level. Check-off the following after inspection:

   A) ____ Sight-down top plate of walls to make sure they are straight:
   ☐ First floor   ☐ Second floor   ☐ Third floor

   B) ____ Plumb all inside and outside corners:
   ☐ First floor   ☐ Second floor   ☐ Third floor

   C) ____ Bolt or shoot-down all bottom plates with approved shots or pins:
   ☐ First floor   ☐ Second floor   ☐ Third floor

   D) ____ Inspect roof sheathing nailing and shear panel nailing so buildings can be wrapped or dried in prior to the framing inspection. Inspect separately if necessary. Make the building inspector aware of shear nailing at roof sheathing inspection. Wrap and weatherproof as soon as possible.

4) ____ Run ceiling joists and check for the following:

     A) ____ Proper size joist for plan:
        ☐ First floor   ☐ Second floor   ☐ Third floor

     B) ____ Proper size and grade of lumber:
        ☐ First floor   ☐ Second floor   ☐ Third floor

     C) ____ Correct span:
        ☐ First floor   ☐ Second floor   ☐ Third floor

     D) ____ Installation of all crowns:
        ☐ First floor   ☐ Second floor   ☐ Third floor

     E) ____ Right size and grade of beams:
        ☐ First floor   ☐ Second floor   ☐ Third floor

     F) ____ Correct connectors, hold-downs, anchor bolts, post anchors, and other hardware:
        ☐ First floor   ☐ Second floor   ☐ Third floor

Plumbing

1) ____ When plumber runs waste, gas, and water, check for the following:

     A) ____ Make sure type and size of pipe is per approved plumbing schematic and plan:
        ☐ First floor   ☐ Second floor   ☐ Third floor

     B) ____ Check for approved sound deadening hangers:
        ☐ First floor   ☐ Second floor   ☐ Third floor

     C) ____ Make sure drains are installed under patio or deck slabs for downspouts prior to pouring:
        ☐ First floor   ☐ Second floor   ☐ Third floor

Framing

1) ____ When installing drop ceilings, look for the following:

     A) ____ Proper size joists:
        ☐ First floor   ☐ Second floor   ☐ Third floor

     B) ____ Proper size and grade of lumber:
        ☐ First floor   ☐ Second floor   ☐ Third floor

     C) ____ Correct span:
        ☐ First floor   ☐ Second floor   ☐ Third floor

     D) ____ Proper connectors:
        ☐ First floor   ☐ Second floor   ☐ Third floor

     E) ____ Installation of all crowns:
        ☐ First floor   ☐ Second floor   ☐ Third floor

2) ____ When installing the subfloor, look for:

     A) ____ The correct type of rated plywood (one-half-inch, five-eighths-inch, three-quarter-inch) and sturdy floor, per plans or specifications:
        ☐ First floor   ☐ Second floor   ☐ Third floor

     B) ____ The correct gluing procedure. *Note*: All subfloors must be glued per plans and manufacturers specifications. The time span between gluing, placement of plywood, and nailing is critical.
        ☐ First floor   ☐ Second floor   ☐ Third floor

C) ____ The proper installation—laid perpendicular to the joists according to the plans:
☐ First floor   ☐ Second floor   ☐ Third floor

D) ____ The correct nailing procedure:
☐ First floor   ☐ Second floor   ☐ Third floor

E) ____ Tongue and groove sheathing:
☐ First floor   ☐ Second floor   ☐ Third floor

Windows and Doors

1) ____ When ordering window sashes and doors, compare the size, type, and quantity called for in the contract and on the plans. Confirm rough opening sizes with framing subcontractor. If all agree, schedule window and door delivery. Check for the following:

A) ____ Proper type, size, style, and finish

B) ____ Correct swing of doors and opening of windows (single- or double-hung, sliding, fixed, or casement)

C) ____ Special or backordered items

Fireplaces

1) ____ Make sure the metal fireplace supplier or mason comes out prior to framing. Confirm with the supplier or framer all floor openings and chimney ties.

2) ____ Check for proper flue lining.

3) ____ Check for proper flue caps and flashings.

4) ____ Schedule fireplace inspection—have the stack accessible.

Roof Rafters and Trusses

1) ____ When stacking roofs or rolling trusses, check for:

A) ____ Proper size and grade of lumber

B) ____ Correct span

C) ____ Correct pitch

D) ____ Proper type of metal connectors

E) ____ Collar ties

F) ____ Roof bracing (purlins, etc.)

G) ____ Ties (seismic ties, struts, and kickers)

H) ____ Blocks (bird blocks, pressure blocks, barge blocks, and chimney blocks)

2) ____ When installing roof sheathing:

A) ____ Check for correct size, rating, and type of plywood.

B) ____ Make sure it is properly laid and blocked. Check whether laps or rafters are specified. Block where necessary.

C) ____ Make sure it is properly nailed with correct size nails. *Note*: Nailing will vary with product.

D) ____ Make sure all vents have one-inch clearance between wood and roofing paper.

E) ____ Check to see if plywood clips are required.

Concrete

1) ____ When chipping for plaster screed:

    A) ____ Make sure perimeter of building and garage are chipped.

    B) ____ Reason they need to be chipped:

        1) ____ Framer measurements are wrong.

        2) ____ Concrete measurements are wrong.

        3) ____ Dimensions on the foundation plan differ from those on the floor plan.

Sheet Metal

1) ____ Run roof flashings, roof-to-wall metal, and valley flashings.

2) ____ Check for correct gravel stop.

3) ____ Check chimney flashings.

4) ____ Make sure all roof vent flashings have been supplied by plumbing, heating, and sheet metal.

5) ____ Install plaster screed or foundation metal:

    A) ____ Check for correct type of metal.

    B) ____ Check for proper lap and nailing.

    C) ____ Check for proper corners.

    D) ____ Check for proper caulking.

6) ____ Make sure overflow scuppers are installed on flat roofs and decks.

7) ____ Make sure downspouts are properly connected to deck drains.

8) ____ Check range hood vent location.

9) ____ Check bath fan housing vent.

10) ____ Check attic ventilation.

Roofing

1) ___ Hot mop flat roofs, decks, stairs, and tubs.

    A) ___ Check contract for special conditions.

    B) ___ Check base flashings, cant strips, and crickets prior to hot mop. *Note:* Positive drainage is critical.

    C) ___ Check how many layers of felt are used and what weight (thirty-, sixty-, or ninety-pound). Make sure the felt is I.C.B.O. approved. Make sure it is installed, nailed, and lapped according to the manufacturer's recommendations.

    D) ___ Check for classification of hot mopping.

    E) ___ Check whether it is to be a fifteen- or twenty-year bondable roof.

    F) ___ Call roofing material manufacturer to review product specifications. Inspect installation of all materials; make sure they are installed according to manufacturer's specifications (lap, nailing, mastic, or windclips).

Windows and Doors

1) ___ Review contract. Confirm who sets window and door frames.

    A) ___ Make sure each window or door is in the right opening.

    B) ___ Check window slide for correct stationary side (XO or OX, single- or double-hung, sliding, or casement).

    C) ___ Make sure window frames are right side up. Make sure weep holes are at the bottom.

    D) ___ Check for correct flashing and proper nailing.

    E) ___ Make sure mullions and grids are installed correctly.

    F) ___ Look for tempered glass where required. *Note:* Make sure the tempered stamp is in the lower corner of windows.

    G) ___ Check doors for proper swing and stationary side.

Wood, Steel, and Wrought Iron Handrails and Railings

1) ___ When installing handrails and railings:

    A) ___ Check for proper material.

    B) ___ Check for correct spacing and heights according to current code.

    C) ___ Check for proper installation.

    D) ___ Check current code for correct balcony railing heights.

    E) ___ Check for proper strength and anchoring.

    F) ___ Check for correct spacing of pickets per plan and code.

Electrical Wiring

1) _____ When installing rough electrical:

    A) _____ Check contract and plans for proper materials and methods.

    B) _____ Make sure electrician has an exterior siding and shear panel schedule so he can set his boxes out accordingly.

    C) _____ Check location of all outlets (television, boxes, and lights).

    D) _____ Check for proper spacing from edge of framing lumber. (Two-inch minimum.)

    E) _____ Check special electrical runs (220-V or 240-V) for dryer, oven, air-conditioning, pool, or spa equipment.

Exterior Siding and Veneer

1) _____ When installing siding or veneer materials, check for the following:

    A) _____ Look for clear seal or primer paint on all exposed wood.

    B) _____ Look for waterproof paper (number fifteen felt), where required. Check for proper lap.

    C) _____ Confirm proper type and texture of materials.

    D) _____ Check for proper nailing and fasteners.

Decking, Framing, and Cleanup

1) _____ Pre-deck coating inspection:

    A) _____ Make sure all holes are plugged.

    B) _____ Make sure all flashings are properly installed.

    C) _____ Make sure crickets are installed prior to preparing decks for waterproofing, to ensure proper drainage.

    D) _____ Check grade and type of preservative used in wood.

    E) _____ Make sure subdeck is dry prior to application of waterproofing materials.

Heating and Air-Conditioning

1) _____ When installing air-conditioning lines·

    A) _____ Check for proper locations of refrigeration and electrical lines.

    B) _____ Check for proper sheet metal and eyebrow vents.

    C) _____ Check for proper insulation.

    D) _____ Check for correct size of line and type.

    E) _____ Check for proper condensate lines. *Note*: Condensate line out-fall must be visible and accessible.

    F) _____ Make sure 220-V or 240-V electrical outlets are prewired to all air-conditioning compressor locations.

2) ____ When installing rough heat, check for the following:
   A) ____ Make sure heating plans have been approved.
   B) ____ Read the contract. Make sure the system being installed is according to the approved plan.
   C) ____ Check for proper size of duct, insulation, and clearances.
   D) ____ Check register locations and return air vent sizes.
   E) ____ Check for proper hangers or straps.
   F) ____ Check for proper clearances of all roof vents from any combustible materials. (One-inch minimum clearance.)
   G) ____ Check for cut or otherwise damaged framing.

Telephone Company

1) ____ During the telephone prewire:
   A) ____ Check for proper locations and quantity.
   B) ____ Check for proper height of outlets, as specified.
   C) ____ Make sure telephone wires are properly run. Make sure they are secured with appropriate slack. Make sure they are wired to the correct locations.

Pre-insulation

1) ____ Pre-insulation is performed prior to insulation, in areas that will later be inaccessible. During pre-insulation:
   A) ____ Check for proper "R" values. Read contract and pre-insulation booklet.
   B) ____ Make sure insulation is properly installed—with the paper facing the heat space.
   C) ____ Make sure it is fastened properly.
   D) ____ Check for insulation behind tubs, shear panels, and attics prior to wrapping (if not accessible after wrapping).
   E) ____ Inspect all pre-insulation before it is covered.
   F) ____ Make sure there are no gaps in the insulation.
   G) ____ Check for complete air infiltration, draft-stopping, or caulking, as required.

Pre-drywall

1) ____ Before hanging the drywall, check for the following:
   A) ____ Check plans for pre-drywall requirements and locations. Make sure required pre-insulation is installed and inspected.
   B) ____ Check contract and plans for any special party or firewall conditions.
   C) ____ Check for proper size, type, and nailing schedule.
   D) ____ Check for proper caulking.

Lightweight Concrete

1) ____ Read the contract.
2) ____ Make sure all holes in subfloors, walls, and decks are plugged.
3) ____ Repair damaged drywall prior to pouring lightweight concrete.
4) ____ Have cleanup subcontractor sweep.
5) ____ Make sure all toilet rings are set at the proper height and backed with three-quarter-inch plywood.
6) ____ Confirm the proper thickness of the lightweight slab.
7) ____ Schedule inspection of subfloor prior to pouring.

Lath-Plaster

1) ____ Before lathing:
   A) ____ Read the contract thoroughly, especially the specification section.
   B) ____ Check for correct gauge of string wire.
   C) ____ Check for furring nails or prefurred lath. *Note*: Properly installed furring nails prevent cracks in the plaster.
   D) ____ Check for waterproof sheathing paper with proper lap (six inches) and shear stapling. Check for proper lap of wire (three inches).
   E) ____ Check for woven wire lath.
   F) ____ Check type of nails and nailing schedule.
   G) ____ Make sure windows, doors, and meter cabinets are flashed prior to lath.
2) ____ Make sure all electrical outlets and door bell wires are not covered after lath installation.
3) ____ Check scaffolding. *Note*: Review safety checklist.
   A) ____ Make sure it is blocked and secured.
   B) ____ Make sure there are two structurally sound planks, side by side, on all runs.
   C) ____ Make sure all scaffolding meets OSHA requirements.

Fireplaces

1) ____ For metal and masonry fireplaces:
   A) ____ Read contract and review plans.
   B) ____ Check for correct firebox size, model, and accessories.
   C) ____ Check for proper drywall installation and taping around fireplace opening.
   D) ____ Make sure clearances and draft blocks are installed properly.
   E) ____ Make sure flue linings and hearth extensions are installed properly.

Plumbing

1) ____ Before installing tubs and showers, check for the following:
    A) ____ Make sure subfloor is cut out for drains.
    B) ____ Check plumber's contract for tub type, size, style, and model number.
    C) ____ Make sure all pre-drywall and pre-insulation are installed.

Cleanup

1) ____ When rat-proofing and fire-stopping:
    A) ____ Make sure cleanup subcontractor does not place any flammable material next to heat vents or water heater vents when draft-stopping or rat-proofing.
    B) ____ When using plasterer's mud, be careful not to allow it to protrude past wall line.
    C) ____ In multi-family or attached units, make sure all debris touching drywall, plaster, or studs at all sound or party wall conditions is removed.

Framing and Masonry

1) ____ Check front and rear patio walls:
    A) ____ Check the architectural plans for locations and details.
    B) ____ Check for weep holes, waterproofing, and french drains as required.

Superintendent's Notes

Walk with municipal inspector on all inspections. Bring the approved plans and specifications. Make sure the subcontractor's foreman walks with the inspector to answer any questions and make necessary corrections for rough inspection approvals.

1) ____ Check when rough heat inspection is complete.
2) ____ Check when rough electrical inspection is complete.
3) ____ Check when rough plumbing inspection is complete.

Prior to the framing inspection, mark floors for corrections:

1) ____ Check wall lines at floor and ceiling.
2) ____ Check window and door opening sizes.
3) ____ Check for special shear blocking.
4) ____ Check joist blocking, spacing, size, and direction.
5) ____ Check all special hardware or framing details.
6) ____ Check nailing schedules.
7) ____ Check all metal connectors.
8) ____ Check plumb of all corners.
9) ____ Check closet door opening sizes (width and height).

10) ____ Check stairs for stringer size and number, plus clearance for skirt boards and drywall. Check for proper stair stringer nailing and blocking. Also, check tread depth and width and riser height.

11) ____ Check medicine cabinet heights and locations.

12) ____ Check tub and shower backing. Check flat block backing for shower and tub jambs.

13) ____ Check cabinet backing. Make sure there is flat block backing in ceiling for upper island cabinets.

14) ____ Check towel bar, toilet paper, and towel ring backing. (Towel bars, fifty-four inches high; toilet paper holders, eighteen inches high; and towel rings, sixty inches high.) Use two-by-six and two-by-eight flat blocks. Mark floors for future identification of backing locations.

15) ____ Check size of all light wells and level of all corners.

16) ____ Check water heater platform size and height. Make sure it is sealed and blocked properly. Make sure all platforms are square, level, and plumb.

17) ____ Check all dropped ceilings for the correct height, width, depth, and proper joist size and grade.

18) ____ Check all post-to-beam connectors.

19) ____ Make sure all window headers are furred level for drywall. Make sure the margins are even. Make sure windows are set square and even to openings.

20) ____ Check all ceilings with an attic above to make sure they have the following:

    A) ____ Attic access

    B) ____ Cat walks

    C) ____ Strong backs

21) ____ Check for all lath or exterior siding backing.

22) ____ Check fireplace faces for drywall backing.

23) ____ Check for all handrail backing (flat blocks).

24) ____ Make sure all plumbing straps and heating straps align, are flush to the wall, and do not bulge drywall.

25) ____ If there is a pair of windows or a door and a window side by side, make sure headers line up. Align properly inside and outside.

26) ____ Make sure there are no extra anchor bolts sticking up in doorways or outside—under patio door sills or mudsills. *Note:* This should have already been checked and corrected. If not, use a piece of pipe and break them off yourself.

27) ____ Make sure water service risers are plumb.

28) ____ Have all staples that held temporary braces removed from exterior wood door jambs.

29) ____ Check sizes of all roof gable vents, foundation vents, storage, and heater room vents.

30) ____ Make sure all exterior wood frames are back primed.

31) ____ Check garage door jambs for proper nailing and grade of lumber.

32) ____ Make sure all window and door frames are properly set and flashed.

33) ____ Make sure all exterior columns have proper size trim applications and are plumb.

34) ____ Check all patio or garden walls for correct height, length, and bracing.

35) ____ Check gas meter vaults.

36) _____ Make sure all shower door openings are plumb.

37) _____ Nail all squeaky floors. Walk and mark areas that need correction.

38) _____ Check elevations for pop-outs around windows and doors. Check elevations for proper veneer, siding, or plant-on details.

39) _____ Check draft-stops and furnace platforms. *Note*: Block or draft-stop as required.

40) _____ Check shear blocking for shear panel and drywall. *Note*: Block framing for shear nailing drywall, as required according to plans.

41) _____ Check nailing of king studs to headers.

42) _____ Check nailing of corners, channels, and intersecting walls.

43) _____ Check fire blocking at dropped ceilings, soffits, and stairs.

44) _____ Check exterior wall lines for straightness and plumb.

45) _____ Check starter board above entries and overhangs for damaged wood, shinners, knotholes, and proper blocking.

46) _____ Check for damaged rafter tails.

47) _____ Check for damaged bird-stops, pressure blocks, barge or fascia blocks, and missing blocks.

48) _____ Make sure backing and blocking is installed and properly nailed.

Garage Doors

1) _____ Before installing garage doors:

A) _____ Check the contract and plans for the correct design and materials.

B) _____ Check for correct hardware and any damage to tracks or runners.

C) _____ Make sure it is built according to materials list on contract, plan elevation, and specifications.

D) _____ Check for proper nailing and hardware.

E) _____ Check for proper clearances.

F) _____ Make sure doors open and close properly.

G) _____ Make sure all shinners are trimmed and clipped flush.

H) _____ Make sure clear seal or primer paint is applied as soon as possible.

5

Finish Trades Production

Introduction

With the majority of structural and mechanical components of the building in place and inspected, the rough phase of construction concludes and the finish phase begins. Quality control during the finish operation is very important because the majority of finish trade materials will remain exposed and forever on display for critical review.

During the finish phase of construction, it is very easy for superintendents and subcontractors to get lulled into complacency because the buildings *appear* to be complete. This is not the time for a project superintendent to lose control of the schedule or product quality.

During the finish phase of construction there are many concurrent subtrade operations taking place. With increased construction activities and shorter completion times allotted to tasks, the project superintendent's work load increases dramatically, particularly in the area of quality control.

Through organization and time management, the project can easily be controlled. In order to withstand the constant pressure to meet completion dates, yet build a quality product, the project superintendent must be organized. Here again, proper planning is the key. Prior to organization, there must be direction. A *realistic* schedule provides this direction. The schedule will not only afford the opportunity to organize the day and week, but the future as well. Remember, the scheduled tasks taking place today will affect what happens tomorrow, next week, even next month. Timely completion of minor tasks will determine whether the project is completed on schedule.

The finish checklist and schedule will enable project superintendents to get an overview of the many simultaneous subtrade activities, and coordinate, monitor, and control product quality. If used on a *daily* basis, this checklist will become a powerful organizational tool. With a good, strong, honest effort, and a little planning, not only can running a successful project be profitable, but it can also be fun! This checklist and your schedule are guidelines, and they must at all times remain flexible and ready to adapt to sudden, unplanned changes.

Finish Trades Checklist

Insulation

1) _____ When installing insulation:

    A) _____ Read the contract.

    B) _____ Check for correct "R" values and locations according to plans and contract.

    C) _____ Check for sound deadened areas.

    D) _____ Make sure insulation is installed according to manufacturer's specifications.

    E) _____ Check for foam around all door and window openings and bottom plates, according to applicable code requirements.

    F) _____ Obtain inspection and certification prior to drywall installation.

Windows and Doors

1) _____ When installing glass:

    A) _____ Read contract and plans.

    B) _____ Check for proper type of finish and hardware:

        Frames: wood_____   metal_____   vinyl_____

        Finish: raw_____   primed_____   painted_____

        Paint: electrostatic (bronze)_____   anodized___

    C) _____ Check for correct series. Make sure grids have the proper cut-up. Make sure grids are not defective or damaged.

    D) _____ Check glazing schedule for special requirements such as type of wind zone or fire rating.

    E) _____ Check for tempered glass. Make sure tempered stamps are in lower corners of windows. *Note*: This is the correct way to glaze glass panels *and* a FHA/VA requirement.

    F) _____ Check for insulated glass or dual glazing.

    G) _____ Make sure screen material is per contract. Check for broken, cracked, missing, or backordered glass as soon as it is installed.

    H) _____ Make sure windows move freely and lock properly.

Drywall

1) ____ When installing resilient drywall channels:
- A) ____ Check contract for correct channels.
- B) ____ Make sure screws and nails are per code and plans.
- C) ____ Check installation per code, contract, and manufacturer's specifications.

2) ____ When installing drywall:
- A) ____ Check contract and plans for special conditions such as party walls, special caulking, and nailing or screwing requirements.
- B) ____ Check type ("X," "CD," regular, or water resistant) and size (one-half-inch or five-eighths-inch).
- C) ____ Check corner bead for proper attachment and plumb.
- D) ____ After drywall is stocked, make sure window panels and mullions have been reinstalled properly and no damage has been done to doors or windows.
- E) ____ Make sure medicine cabinets, outlets, and attic accesses are not covered.

3) ____ During the lath inspection:
- A) ____ Check plans and contract for special conditions, such as expansion joints, control joints, waterproofing, and fireplace faces.
- B) ____ Check type of paper, lath, and wire.
- C) ____ Check for correct type and length of nails.
- D) ____ Check for proper corner bead and nailing.
- E) ____ Check tie wire for proper spacing and tightness.
- F) ____ Check for covered lights and outlets.
- G) ____ Patch any holes in lathing paper.

4) ____ During the drywall inspection:
- A) ____ Check for correct nailing procedure, per plan.
- B) ____ Make sure window and door frames are lapped over, with no joints or sheet breaks. Check for a header or king stud connection.
- C) ____ Check for proper caulking and sealing.
- D) ____ Check for proper corner bead attachment.
- E) ____ Make sure drywall is no more than one-half inch above floor so base board has backing, especially in vaulted ceiling areas.
- F) ____ Make sure no outlets are covered.

Plaster

1) ____ If applying stucco scratch coat:
- A) ____ Check contract for proper type of materials.
- B) ____ Make sure all lath is covered by stucco.
- C) ____ Check specifications in contract for:
  - 1) ____ Scratch coat application

      2) ____ Brown coat application

      3) ____ Finish coat application

      4) ____ Oiling or masking off all glass and frames

      5) ____ Masking off gables and vents

   D) ____ Check for square corners and good, clean reveals.

   E) ____ Cleanup—hose excess from all door and window frames. Dig fall-down excess away from foundation after scratch and brown coats. Clean flat roofs, and hose down.

2) ____ Prior to drywall texture, check the following:

   A) ____ Check contract for materials list.

   B) ____ Check for correct number of drywall mud applications.

   C) ____ Check for correct texture finish (heavy, knock-down, fog, splatter, orange peel, hand trowel, or skip trowel).

   D) ____ Check kitchen and bath applications.

   E) ____ Check for bad taping and nail pops.

   F) ____ Make sure areas to be protected are masked properly.

   G) ____ Demand proper cleanup, including:

      1) ____ Clean out all debris from hanging operations.

      2) ____ Clean floors after texture.

      3) ____ Blade and sand down texture.

      4) ____ Cut-in acoustic ceilings, scribe in ceiling lines.

      5) ____ Clean up, clean up, clean up!!!

3) ____ Inspect brown coat (after seven-day cure time):

   A) ____ Read the contract.

   B) ____ Make sure combined thickness of scratch coat and brown coat is per plans and specifications. Moisten scratch coat prior to brown coat application.

   C) ____ Keep plaster moist so it can cure properly.

   D) ____ Make sure there are no voids in plaster. Make sure finish coat is only one-eighth inch thick. Patch holes or damage prior to color coat.

   E) ____ Check brown coat for hardness by scratching with a 16-D nail. If nail scrapes along the top of the surface, it has cured enough; if it gouges down into plaster, it has not. Check for cracks in brown coat.

   F) ____ Watch for paint over-spray on brown coat that will destroy the bond.

   G) ____ Make sure no outlets have been covered.

Heating and Air-Conditioning

1) ____ Check contract for type and size of furnace or heating unit.
2) ____ Install heating units; set registers.
3) ____ Check for correct finish on registers.
4) ____ Check size, manufacturer, and model number on air-conditioning units.
5) ____ Check thermostat locations and condenser locations.
6) ____ Check location of 220-V or 240-V outlet. Make sure it is not behind nor too close to the condenser unit.
7) ____ Make sure dryer vent is not within six feet of the unit.
8) ____ Check for proper condenser unit location and elevation.

Cabinets

1) ____ When installing cabinets:

A) ____ Check contract for type of wood, materials, and design. *Note*: Certain species of wood are prone to harbor powder post beatles. Check all cabinets prior to installation.

B) ____ Check drawer and door design. (Plastic or metal rollers; raised or recessed panels.)

C) ____ Check hinge type.

D) ____ Check for proper toe-kicks and scribe molding.

E) ____ Check cabinet and counter top heights.

F) ____ Check for full partitions.

G) ____ Check door stiffeners at pantries and broom closets. Replace any warped cabinet doors.

H) ____ Check nailing and alignment of rough top. Make sure holes are cut properly for all sinks and appliances. Make sure sink templates have been supplied by the plumber, and appliance cutouts by the supplier.

I) ____ Check refrigerator, oven, stove, and dishwasher space openings.

J) ____ Check margins at window openings.

K) ____ Check for scribe molding at hall and laundry cabinets.

L) ____ Check alignment of all doors, drawers, and dummy drawers, horizontally and vertically. Check for even margins.

M) ____ Check all drawers for ease of slide. Repair or replace any stiff sliding drawers.

N) ____ Make sure holes have been cut for electrical outlets and kitchen island cabinets.

Finish Carpentry

1) _____ Check plans, interior elevations, and the description of materials in the contract for type, quantity, and quality.
2) _____ Make sure finger joints are all sanded flush.
3) _____ Make sure bases and casings are nailed properly. Make sure all nails are set and filled.
4) _____ Make sure doors are hung properly and are not warped.

Plaster

1) _____ Before applying finish coat, check contract.
2) _____ See color chart for colors.
3) _____ Verify type of finish (heavy texture, lace, or smooth sand finish).
4) _____ Check for covered door bells and electrical outlets.
5) _____ Check for covered light boxes, air-conditioning outlets, porch light boxes, and convenience outlets.
6) _____ Check for cracks, especially around door and window headers, and first and second floor transitions. Fix and patch any defects prior to finish coat application.

Masonry

1) _____ When installing fireplace faces and exterior veneer:
    A) _____ Check contract for materials, and plans for elevation view.
    B) _____ Check for level, plumb, and neatness of joints.
    C) _____ Make sure there are no chipped, damaged, defective, or discolored materials.
    D) _____ Check length, width, and height of hearths.
    E) _____ Check for correct margin all around.
    F) _____ Make sure masonry is brushed clean.

Paint

1) _____ Before finishing cabinets:
    A) _____ Check color chart for type of finish and verify optional colors.
    B) _____ Make sure painter accepts the cabinets as complete and ready for finish.
    C) _____ Check contract for list of materials and required number of coats.
2) _____ After finishing, check for color variations, rough surface, bubbles, or runs in the finish.

Tile

1) _____ When installing ceramic tile around tubs, showers, and sinks, check for the following:
   - A) _____ Check plans and contract for type of tile and colors for various locations.
   - B) _____ Make sure all tile is installed level and plumb. Make sure it has equally grouted joints. Make sure it is smooth and flush.
   - C) _____ Make sure margins are equal and level, with even lines.
   - D) _____ Make sure all cut tile on counter top is ground smooth.
   - E) _____ Trim all nails, excess paper, and grout at v-caps.
   - F) _____ Confirm that green board or walls have been lathed, scratched, and floated.
   - G) _____ Check the float prior to setting tile. Make sure it is plumb, level, square, and true.

Roofs

1) _____ When installing tile roofs:
   - A) _____ Read contract and plans for specifications and locations. Type of roofing:
     Hot mop _____ Composition _____ Shake _____ Tile _____
   - B) _____ Look for proper underlayment. Check lap and nailing of material.
   - C) _____ Make sure all roof jacks, saddles, and flashings are installed, capped, flashed, and glued. Make sure they are painted properly.
   - D) _____ Make sure roofs are complete and roof metal is painted.

Paint

1) _____ Make sure all nails are set and filled.
2) _____ Make sure drywall is checked prior to painting.
3) _____ Make sure all trim joints are tight.
4) _____ Check caulk and prep work in all areas prior to paint or stain.
5) _____ Check for undercoat and primer on all woodwork prior to finish coat.

Cleanup and Framing

1) _____ Make sure floors are sanded.
2) _____ Make sure sander gets into all corners.
3) _____ Check for squeaky floors. Walk and mark all squeaks.
4) _____ Make sure all fractured plywood is replaced.
5) _____ Make sure all holes are filled.
6) _____ Make sure all nails are set.

Weatherstrip

1) _____ When installing weatherstripping:

    A) _____ Check plans for locations.

    B) _____ Check contract for type and finish.

    C) _____ Make sure thresholds are caulked underneath.

    D) _____ Make sure thresholds are screwed down.

    E) _____ Check for damage caused by other trades.

    F) _____ Install all lock-up hardware before installing weatherstrip.

Counter Tops

1) _____ When installing counter tops:

    A) _____ Read contract and manufacturer's specifications. Confirm color with contract.

    B) _____ Check type of counter:
Formica _____ Cultured marble _____ Onyx _____ Corian _____

    C) _____ Make sure the correct size bowl and set back are cut. (Use a bowl or sink template from the plumber.) Also, provide the faucet hole and size layout.

    D) _____ Make sure the bowl is in the proper location and does not interfere with drawers.

    E) _____ Make sure counter top has proper overhang or edging.

    F) _____ Check splashes—make sure they are cut properly.

    G) _____ Check caulking—make sure it is neat and complete.

Appliances

1) _____ Upon delivery of appliances:

    A) _____ Read contract to confirm cutout sizes.

    B) _____ Check colors.

    C) _____ Check for upgrades and standards.

    D) _____ Write down all serial numbers. File with a date stamp.

    E) _____ Check for proper fit in cutouts and even reveals.

Heating and Air-Conditioning

1) _____ Before setting air-conditioning units, after finish grade:

    A) _____ Read contract and approved plans.

    B) _____ Check size of units and confirm manufacturer.

    C) _____ Verify clearances and locations.

    D) _____ Check electrical hook-ups (220-V or 240-V).

    E) _____ Check for pre-cast concrete or poured-in-place pads for condensers:

        1) _____ Pre-cast should be supplied by air-conditioning subcontractor.

        2) _____ Poured-in-place should be supplied by concrete subcontractor.

Finish

1) _____ When installing finish hardware:

    A) _____ Check contract for quantity and finish.

    B) _____ Check all doors, especially entry doors, for over-bores or damage caused by the hardware installer.

    C) _____ Verify and sign for materials. Double-check quantity against field changes.

    D) _____ Check bath accessories such as towel bars and toilet paper holders. Make sure they are fastened and secured to the walls properly.

Cleanup

1) _____ Clean windows.

2) _____ Make sure window cleaner does not use abrasives.

3) _____ Make sure cleaning agent is diluted so it does not damage the exterior windows or frames.

4) _____ Check putty knives for nicks and burrs to prevent scratching.

5) _____ Make sure window cleaner cleans up spilled water immediately, especially on the second floor. *Note*: The more soap and water used, the less scratching to glass. Have the plastering contractor power spray the windows as soon as possible after the finish coat; this will greatly reduce scratched glass. Masking off all doors and windows prior to plaster is another excellent way of preventing scratched glass.

Electricity

1) _____ Before setting finish electric:

    A) _____ Read contract for list of materials and specifications.

    B) _____ Get all testing data from electrical subcontractor for fire alarm system, including fire department releases.

    C) _____ Keep a broken fixture list for every day that the electrician hangs fixtures. Insist on a daily damage report. File with the contract.

    D) _____ Make sure the electrician "rings-out" each building as he sets trim. Do not wait until final inspection.

    E) _____ Make sure the electric panel breakers are clearly marked and identified by the electrician. Do not use abbreviations.

Deck Coating

1) _____ During final deck coating:
  A) _____ Read contract for quantity and type of finish material and applications.
  B) _____ Check for proper drainage, crickets, and lightweight concrete build-ups, if necessary, to ensure proper drainage. *Note*: This must be done prior to the application of final deck coatings.
  C) _____ Do a water test upon completion.

Luminous Ceilings

1) _____ Before installing attic scuttles, light wells, and luminous ceiling frames, read contract for list of materials and type of finish.
2) _____ Make sure all miter joints are closed up. Make sure ceiling fits opening with no voids. Make sure miters are tight.
3) _____ Check for correct installation.
4) _____ Make sure finish is not damaged.
5) _____ Make sure plastic panels are not damaged.
6) _____ Caulk any gaps.

Plumbing

1) _____ When setting finish plumbing:
  A) _____ Read contract specifications and materials list.
  B) _____ Check installation on contract.
  C) _____ Make sure there are no wrench marks or kinks on finish trim or pipes.
  D) _____ Check under sinks for leaks. Check toilets for proper flush and tank fill.

Windows and Doors

1) _____ Make sure screen frames are tight at mitered corners.
2) _____ Make sure there are no rips in material or dents in frames.
3) _____ Make sure screens fit properly and are easily removed.
4) _____ Make sure sash moves and latches properly.
5) _____ Make sure track is cleaned out.
6) _____ Make sure slide and lock adjustments are complete.
7) _____ Replace any damaged locks and hardware.

Superintendent's Note

Notify utility companies three to four weeks prior to final production so they can schedule and complete their make-up. You should personally contact gas and public service companies. This may eliminate any last minute surprises.

Canvas

1) _____ Install cloth awnings (if applicable).
2) _____ Check for correct color combination and pattern.
3) _____ Check for correct hardware according to contract.
4) _____ Make sure any holes drilled in exterior are properly sealed and caulked.

Shower Doors

1) _____ Before installing shower doors, check the following:

    A) _____ Read the contract.
    B) _____ Check type of finish and material:
         Chrome _____ Brass _____ Antique _____
    C) _____ Check type of glass:
         Clear _____ Obscure _____
    D) _____ Check door latch adjustment. Make sure top track is properly secured.
    E) _____ Check door drip adjustment.
    F) _____ Check waterproof vinyl flap and drip extrusions.
    G) _____ Make sure doors are properly and neatly caulked with silicone or approved sealant.

Mirrors and Glass

1) _____ Read contract and plans.
2) _____ Check channels and finish. Make sure all edges are polished, not seamed.
3) _____ Check mirror color, type, finish, and thickness (one-quarter-inch, three-sixteenths-inch).
4) _____ Check wall and ceiling reveals.
5) _____ Check for chips on all four corners.
6) _____ Check for black spots, scratches, cat-eyes, chips, and other defects.
7) _____ Check for correct size and location.
8) _____ Check spacing and types of clips. *Note*: FHA/VA requires double clips.
9) _____ Make sure the mirror channel is properly cut—not too long to cause a safety hazard from a sharp edge on the aluminum channel.
10) _____ When installing mirrored closet doors:

    A) _____ Read contract for description and installation procedures.
    B) _____ Make sure openings are square and doors slide against bumper jambs properly.
    C) _____ Check top and bottom tracks. Make sure they are cut square and tight to jambs.
    D) _____ Check for defects and distortions.

Floor Coverings

1) ____ When laying entry vinyl, ceramic tile, or wood parquet flooring, check the following:

    A) ____ Read contract to confirm materials and installation procedures.

    B) ____ Make sure it is laid square to walls and openings. Make sure it has an even margin at walls.

    C) ____ Make sure step downs have finished edges.

    D) ____ Confirm correct grout color, floor finish, or pattern.

    E) ____ Check for defects or discolorations in grout, floor finish, or pattern.

    F) ____ Protect flooring with cardboard or visqueen immediately after installation.

    G) ____ Reinstall baseboard or base shoe, as required.

2) ____ When installing resilient flooring:

    A) ____ Check floor plan, flooring contract, or homeowner selections for correct pattern.

    B) ____ Make sure material is cut tight to walls with no obvious seams.

    C) ____ Make sure vinyl flooring is caulked at tubs, showers, and water closets.

    D) ____ Check for damage at kitchen appliances.

    E) ____ Roll vinyl to eliminate air bubbles.

    F) ____ Check manufacturer's specifications for proper seam sealer.

3) ____ When installing carpet:

    A) ____ Check with flooring contractor to verify type, style, manufacturer, quality, and color of carpet.

    B) ____ Make sure tack strip is nailed.

    C) ____ Check for right type of pad and glue.

    D) ____ Make sure seams are correct and trimmed.

    E) ____ Make sure carpet is stretched.

    F) ____ Make sure carpet layers do not take out or damage window screens when they run their power cords into the house.

    G) ____ Make sure any doors removed by flooring subcontractor are replaced.

    H) ____ Make sure flooring subcontractor notifies superintendent if doors need to be cut down.

    I) ____ Check all doors immediately after carpet is laid to determine if any doors need to be cut down.

    J) ____ Check electric meters. *Note*: Jumping meters is against the law! If the temporary power has been disconnected, the carpet installers must provide their own power through the use of portable generators.

6

Final Production

Introduction

This final production checklist is designed to "complete" and fine tune the house, inside and out. It takes the superintendent around the entire exterior of the building, through the garage, and into every room, closet, and cabinet. The concept is to go around the building the very same way you would when making a materials take-off. Basically, it is an exercise in observation, notation, and follow-through.

The problems are similar to those that exist in framing. After you've checked a few houses, you will find that the same subtrades make the same mistakes, over and over again. Once you identify these patterns, it will become easier to investigate and correct the problems.

Although walking and writing up corrective tickets in houses can be extremely monotonous and time consuming, it must be done on a *daily* basis. Usually, the final checklist will take about the same amount of time it takes to mark out final framing pick-up, that is, one to two hours per house. With this time frame, and your homeowner occupancy dates in mind, it is sometimes necessary to work overtime in order to keep ahead of the move-ins.

By going through the house, writing up corrective tickets, and sending the final wave of subtrades through the house to make the necessary repairs, you will not only increase the standards of quality, but eliminate many surprises. In the end, your company will be able to proudly deliver a quality product to the homebuyer.

Final Checklist

Exteriors

The building exterior may be stucco, brick, block, stone, siding (wood, vinyl, or aluminum), hardboard, or wood (clapboards, shingles, T1-11, board and batten, or rough sawn lumber).

1) _____ Roll out plans and study exterior elevations. *Note*: This is a final check.

2) _____ Compare exterior elevations and plans for proper design and correct position of all plant-ons.

3) _____ Study joint trench utility plans prior to digging for mail boxes.

4) _____ Repair or replace any defective or knotholed plant-ons.

5) _____ Repair or replace any damaged starter board. Caulk or fill any sizeable knotholes.

6) _____ Check for shinners in starter board, especially above entry.

7) _____ Make sure driveway is according to plan, with control joints and no damage from fine grading.

8) _____ Check for future irrigation sleeves under all driveways *prior to pour*.

9) _____ Check entry walk and front porch area for defects, cracks, or damage in concrete.

10) _____ Make sure all trash is removed and lots cleaned prior to finish grading. Remove all roofing debris.

11) _____ Make sure lot is graded per precise grading plan. Make sure lots drain properly away from the house. Make sure swales are cut properly to area drains.

12) _____ Make sure air-conditioner condenser pad is set.

13) _____ Check for proper condensate disposal.

14) _____ Check electrical connections to air-conditioner. Make sure air-conditioner disconnect is correct and not damaged or missing. Make sure fuse at disconnect is properly labeled by electrician—*do not pull under load*.

15) _____ Make sure electric meter, telephone, and television are properly grounded, with the ground clamp visible. Make sure breakers are clearly labeled.

16) _____ Make sure all lids or covers to utility cans are securely fastened—not missing or damaged.

17) _____ Make sure all rain gutters are complete, properly secured, painted, and tied into area drains. *Note*: Make sure splash blocks are in place if rain gutters are not tied into area drain system.

18) _____ Check for sheet metal rain diverters on roof above entry doors. *Note*: This is a FHA/VA requirement.

19) _____ Make sure all roof vent caps and storm collars are installed and painted.

20) _____ Make sure roof vents are extended high enough per code for final inspection.

21) _____ Make sure sewer clean-out is exposed approximately two inches above grade.

22) _____ Make sure air-conditioning condensate line is stubbed out the side of the building and unplugged, with proper disposal pit.

23) _____ Make sure eyebrow vent at air-conditioner is packed with insulation or foam.

24) ____ Make sure any waterproof electrical outlets and fixtures have not been covered by exterior wallcovering.

25) ____ Check all window frame weep holes. Make sure they are free of caulk and debris.

26) ____ When walking around the building, list any cracked or damaged exterior wall-coverings as well as any damaged or missing roofing material.

27) ____ Check exterior materials for proper nailing and damage.

28) ____ If stucco, check for "equator" cracks between first and second floors at the rim joist.

29) ____ Check color and texture of exterior for full coverage and uniform pattern.

30) ____ Check for discolorations in the exterior. Make sure repairs are properly blended. Waterproof a minimum of three feet around the entire building as soon as possible after repairs and blending are complete.

31) ____ Check for voids in exterior at recessed light fixtures, especially above the entry. Make sure fixture trim covers cutout.

32) ____ Make sure scratch and brown coats were dug back far enough away from the building to ensure even and full coverage, especially below door thresholds.

33) ____ Make sure gas meter stub and building stub are aligned and in correct locations.

34) ____ Make sure all hose bib handles are installed and not damaged. Make sure they are painted to match the exterior. (Same color as sheet metal.)

35) ____ Prior to paint, make sure all surfaces have been properly cleaned and primed.

36) ____ Check plans for height of fireplace stack. Check all high and low profile aluminum chimney caps. Make sure they are properly installed and secured.

37) ____ Check masonry for damaged or missing bricks. Repair or regrout any loose bricks, voids, or "bee-holes."

38) ____ Wash or sand blast all masonry to remove mortar, paint, or caulk, if necessary.

39) ____ Make sure all exterior light fixtures are installed.

40) ____ Make sure dryer vent is clear. Make sure the damper works properly. Make sure it is six feet away from air-conditioning unit.

41) ____ Make sure condensate line drains properly and is clear of debris.

42) ____ Make sure all exterior patches are complete and painted.

43) ____ Make sure all chimneys are sealed and stucco chimneys are waterproofed.

44) ____ Make sure fascia or barge blocks at chimney are not missing. Make sure they arc properly blocked and nailed.

45) ____ Make sure chimney is counter-flashed and painted.

46) ____ Make sure all flashings are painted.

47) ____ Make sure all exterior paint is complete. Check all exterior woodwork, starter board, eves, and overhangs.

48) ____ Make sure all exterior wood siding (if applicable) has been nailed and painted.

49) ____ Check the installation of all door thresholds and weatherstripping. Make sure they have not been damaged.

50) ____ Make sure screens have been installed and not damaged.

51) ____ Make sure electric panel has all circuit breakers identified. Make sure wallcoverings do not impede closing or fastening of lids.

52) ____ Make sure any missing or stolen light fixtures have a blank plate or fixture prior to final inspection. Make sure exposed, bare wires are capped, with the outlet blanked off.

53) _____ Make sure address numbers are complete, level, and per governing agency requirements.

54) _____ Check for broken or missing windows. Make sure all broken glass is replaced.

55) _____ Check doors and window frames for dents and damage. Replace any missing mullions, mul-bars, or screws in mul-bars.

56) _____ When roof is complete, remove excess materials and debris. Replace any broken or missing materials. Make sure valleys are cleaned, and flashings, vent caps, and storm collars are installed and painted.

57) _____ Check doors and frames for damage. Make sure they slide, lock, and latch properly.

Electric Panel

1) _____ Make sure all covers are installed and secured.

   A) _____ Check when electric panel is made up and complete.

   B) _____ Check when telephone can is made up and complete.

   C) _____ Check when television can is made up and complete.

2) _____ Make sure all circuit breakers are clearly marked.

3) _____ If meter closet is a combination of gas and electric, make sure:

   A) _____ The door closes and latches properly.

   B) _____ The door is not warped, delaminated, or damaged.

   C) _____ The door is finished on all six sides.

   D) _____ The metal cap and drip are installed on the top rail of doors.

   E) _____ There is a two-inch P.V.C. sleeve under the door for gas meter stub. Make sure the sleeve is open, with no obstructions.

   F) _____ The inside of the cabinet is completely drywalled with five-eighths-inch, type "X" drywall. Make sure it is taped, with the lower half sealed. Caulk all gaps and voids.

   G) _____ The shelf separating the gas and electric is weatherstripped and draft-stopped.

   H) _____ The door is vented and louvered, with the bottom vents open and the top covered.

   I) _____ When properly sealed-off at the shelf, it separates gas and electric and is legal per code. *Note*: Check with the local public utility and build to approved specifications.

Decks and Balconies

1) _____ Check doors leading to decks and balconies for easy movement and proper lock.

    A) _____ If french doors—make sure they are completely painted. Make sure thresholds and hardware are complete and not damaged.

    B) _____ If aluminum sliding doors—make sure handles and bumpers are complete. Make sure the screen slides and locks easily and is free of defects. Check the door sill for damage.

2) _____ Make sure deck coating is sanded and complete. Check for scrapes and damage. *Note*: Do not sand pressure treated lumber.

3) _____ Check for clogged deck drain and scupper. Do a water test prior to move-in.

4) _____ Make sure flashing and counter-flashing are complete, painted, and not damaged.

5) _____ Make sure both ends of handrail are firmly secured.

6) _____ Check paint on handrail (if applicable).

7) _____ Make sure downspouts from drains have been installed and painted.

8) _____ Make sure exterior lights are installed and not damaged.

9) _____ Check all waterproof electrical outlets.

10) _____ Check roof overhang for shinners. Clip all shinners flush and touch up.

11) _____ Check exterior for cracks or damage.

12) _____ Check exterior for water stains and paint or blend, as required.

13) _____ Make sure door thresholds have been properly caulked and weatherstripped.

Garage

1) _____ Check plans for the correct plant-ons on all garage doors.

2) _____ Make sure garage doors are hung properly (square to the opening) with an even reveal all around.

3) _____ Check garage door jambs for splits and defects. Make sure garage door hasp is not missing and bore hole lines up. *Note*: Hasps tend to get bent when left in the open position by painters while painting doors.

4) _____ Check for loose lag bolts attached to garage door frame.

5) _____ Check for gaps between garage door jambs and stem wall.

6) _____ Check the backside of garage door for shinners. Clip any shinners.

7) _____ Check garage slab for expansion joints, saw cuts, or quick joints.

8) _____ Saw cut driveway the day after pour.

9) _____ If furnace, or heating unit, and water heater platforms are in the garage, make sure bumper pipes are installed, painted, and capped per code. *Note*: Gas pilots must be eighteen inches above floor when in garages.

10) _____ Check water heater "pop-off" (T/P valve). Make sure it is properly secured and not clogged. Make sure it works properly with no leaks, and terminates within six to eight inches of the floor.

11) _____ Make sure electrical chase is drywalled and properly blocked or draft-sealed.

12) _____ Check heating unit or water heater vent for proper clearance at floor or roof sheathing. *Note*: Minimum one-inch clearance around vents.

13) _____ Make sure any holes in firewalls are sealed. Check drywall patching, draft-stops, and blocks.

14) _____ Make sure water heater platform is level prior to installing heaters.

15) _____ Make sure heating unit platform is level prior to installing units.

16) _____ Make sure building card is accessible, and insulation and roofing are certified. Keep all certifications in the jobsite filing cabinet.

17) _____ Make sure baseboard is installed behind the heating unit and water heater.

18) _____ Make sure any exposed sheet metal vent pipe is properly secured with three self-tapping sheet metal screws.

19) _____ Make sure dual-wall heat vent is properly lapped and securely screwed.

20) _____ Check for plastic fittings sticking through firewall. *Note*: This is a code violation.

21) _____ Check all base screed vents or "suicide" vents in garage. Make sure vents have not been covered by drywall, siding, or plaster.

22) _____ Check garage side-yard door (if applicable) for correct swing. Make sure door closes properly and is not warped or delaminated. Make sure its hardware is complete and locks and latches operate properly.

23) _____ Remove any lumber or trash at garage rafters.

24) _____ Clean out garages.

25) _____ Patch any holes in drywall, especially on the firewall.

Interiors

Entry Area

1) _____ Make sure entry fixture is not broken or missing.

2) _____ Check baseboard and base line for bows in the walls.

3) _____ Check for acoustic ceiling scrapes and wall scrapes.

4) _____ Check front door for:

    A) _____ Correct swing

    B) _____ Complete hardware and easy latch

    C) _____ Correct door design and quality (*Note*: Check the number and design of door panels.)

    D) _____ Cracks, pitch pockets, damage, and other defects

5) _____ Cross-sight door. String line if necessary to check for "dog-legged" door jambs.

6) _____ Check door jambs and casings for damage. Make sure all hinge pins have been knocked down.

7) _____ Look for flexed or rigid door stops.

8) _____ Check finish on top and bottom of door. Make sure all six sides have been finished or sealed.

9) _____ Check floor coverings (as required):

    A) _____ Make sure carpet is complete, properly stretched, and trimmed, with no bad seams.

    B) _____ Make sure ceramic tile is complete and grouted.

        C) _____ Make sure wood parquet floors have been installed completely and protected.

        D) _____ Make sure base shoe is complete and painted as required.

         Note: Make sure flooring subcontractor protects the floors by covering with cardboard, paper, or visqueen.

10) _____ Check sidelight glass for the correct type.

11) _____ Make sure sidelight frame is not damaged or bent.

12) _____ Make sure there are no missing window stops.

13) _____ Make sure sidelight glass is not broken or pressure cracked in the corners.

14) _____ Check entry closet appearance: Good _____ Bad _____ .

15) _____ Check inside entry closet for damaged drywall, complete baseboard, level shelf and pole, wiring of doorbell sending-unit (if applicable), damaged or missing light fixture, and complete paint.

16) _____ Make sure entry railing (if applicable) is installed and properly secured. Check for any missing, bent, or broken items.

17) _____ Check the drywall for:

        A) _____ Nail pops

        B) _____ Bad tape joints

        C) _____ Damaged corner bead

        D) _____ Straight lines at off-angles

        E) _____ Poor or light acoustic coverage

        F) _____ Scrapes in ceiling and walls

        G) _____ Fractures, tears, or hammer holes

        H) _____ Blade marks and rough or uneven texture pattern

        I) _____ Properly sanded walls and cut-in ceiling

        J) _____ Full and complete paint coverage

Living Room

1) _____ Check baseboard and base line for straight alignment and bows in the walls. Replace any damaged or missing baseboard. Check for missing or damaged door casings.

2) _____ Check drywall. Patch and texture as required.

3) _____ Make sure all windows are clean. Replace any broken glass.

4) _____ Make sure all windows slide freely and latch properly.

5) _____ Make sure screens fit properly, with no bent frames, tears, or damage to screen material.

6) _____ Make sure all heat registers are caulked, finished, level, and complete.

7) _____ Make sure return hot-air grill is finished.

8) _____ Make sure all electrical outlets and switch plate covers are installed level. Look for cracked or broken face plates and for pressure cracks at screw heads.

9) _____ Make sure window stools and aprons are free of defects, dents, splits, and hammer tracks. Make sure all nails have been set, filled, and properly caulked.

10) _____ Make sure all exposed beams are sandblasted, stained, and in good shape.

11) _____ Check for even margins on the tops and sides of all windows.

12) _____ Check wrought iron and wood handrails for secure installation and proper finish, if applicable.

13) _____ Make sure fireplace doors are in place and adjusted.

14) _____ Make sure the hearth is clean.

15) _____ Make sure the masonry is clean.

16) _____ Make sure all brick grout joints are clean, and any voids or "bee holes" are filled. Make sure all grout has been properly pointed up.

17) _____ Check fireplace damper; make sure it works freely. Make sure fireplace screens or glass doors have been installed free of damage.

18) _____ Make sure mason fills any holes in the firebox and replaces any cracked firebox bricks.

19) _____ Check for draft-stops as required by building code when using metal fireplaces.

20) _____ Make sure log lighter trim is complete with chrome valve cap, valve handle, and key (if fuel gas is supplied).

21) _____ Check floor coverings (as required):

    A) _____ Make sure carpet is complete, properly stretched, and trimmed, with no bad seams.

    B) _____ Make sure ceramic tile is complete and grouted.

    C) _____ Make sure wood parquet floors have been installed completely and protected.

    D) _____ Make sure base shoe is complete and painted as required.

22) _____ Check for full and complete paint coverage.

Dining Room

1) _____ Check base line for wall bows.

2) _____ Make sure all baseboard is complete and any damaged or missing base has been replaced. Check door casings for damage.

3) _____ Make sure all windows slide freely and latch properly. Check frames for damage.

4) _____ Check corners of windows for pressure cracks.

5) _____ Check drywall for even margins at tops and sides of all windows.

6) _____ Make sure screens fit and are damage free.

7) _____ Make sure window stools and aprons are free of defects, and are finished according to specifications.

8) _____ Make sure patio doors and screens (if any) are installed. Make sure doors move easily and latch properly.

9) _____ Make sure heat registers are caulked, finished, and installed level.

10) _____ Make sure return hot-air grill is level and painted.

11) _____ Make sure electrical outlets and switch plate covers are level and flush to the walls, with no pressure cracks at screw holes.

12) _____ Check drywall for:

    A) _____ Nail pops

    B) _____ Bad tape joints

    C) _____ Damaged corner bead

    D) _____ Straight lines at off-angles

E) ____ Poor or light acoustic coverage

F) ____ Scrapes in ceiling and walls

G) ____ Fractures, tears, or hammer holes

H) ____ Blade marks and rough or uneven texture pattern

I) ____ Properly sanded walls and cut-in ceiling

J) ____ Full and complete paint coverage

13) ____ Make sure light fixtures are not damaged or missing.

14) ____ Check floor coverings (as required):

A) ____ Make sure carpet is complete, properly stretched, and trimmed, with no bad seams.

B) ____ Make sure ceramic tile is complete and grouted.

C) ____ Make sure wood parquet floors have been installed completely and protected.

D) ____ Make sure base shoe is complete and painted as required.

15) ____ Check coffered ceiling (if applicable). *Note*: Look for clean, straight drywall lines at off-angles.

Kitchen

1) ____ Make sure cabinets are complete. Check the overall appearance of cabinets: Good _____ Bad _____ .

2) ____ Check cabinets for smooth, even finish.

3) ____ Make sure scribe molding is complete on sides and bottoms of upper cabinets.

4) ____ Make sure toe-kicks are complete.

5) ____ Replace any warped cabinet doors.

6) ____ Check all door and drawer alignments. Make sure the tops of all doors are in alignment, horizontally and vertically, with even margins.

7) ____ Adjust any stiff sliding drawers. Check for damaged rollers and drawer guides.

8) ____ Check shelf cleats inside cabinets for proper nailing; make sure there are no bent nails in cabinet cleats.

9) ____ Make sure shelves have been nailed or stapled properly. Look for missed nails through pressed wood shelves.

10) ____ Replace any sprung hinges and make any necessary hinge adjustments.

11) ____ Check bread board installation (if applicable).

12) ____ Look for door stiffeners at all pantry doors. Make sure pantry shelves have back-up blocks under all shelves. Make sure they are attached to the face frames.

13) ____ Make sure all holes have been cut for electric and plumbing.

14) ____ Make sure walls inside cabinets have been finished appropriately.

15) ____ Check for missing, broken, or defective ceramic tile (if applicable). Check for missing v-caps, beaks, or one-quarter rounds; make sure all pieces are complete, damage-free, and grouted.

16) ____ Make sure tile has been grouted properly.

17) ____ Trim excess paper and grout behind or under v-caps, especially under the sink.

18) ____ Check for shinners, especially under the sink and counter tops.

19) ____ Check kitchen sink for defects, chips, scratches, dents, dings, cat-eyes, and any other construction related damage.

20) ____ Make sure garbage disposal has been installed on the dishwasher side of the sink. Make sure the plug or pigtail reaches the electrical outlet, with no kinks in any of the plumbing lines, especially in the dishwasher line.

21) ____ Make sure the dishwasher has been installed and the plug or pigtail reaches the electrical outlet.

22) ____ Make sure the dishwasher is level. Make sure the legs have been adjusted. Make sure it has been set square, evenly aligned to the cabinet opening, and securely screwed to the cabinet face frame.

23) ____ Make sure the range has been installed, set square, evenly aligned to cabinet opening, and securely screwed to the cabinet face frame. *Note:* Check for customer upgrades. Do not install cooktop inserts until walk-through.

24) ____ Make sure the range hood has been installed and not damaged. Make sure the range vent and damper have been installed, with the vent properly duct taped. *Note:* Make sure the vent does not cover the 110-V electrical outlet.

25) ____ Check oven with upgrade list. Make sure all homeowner upgrades have been installed.

26) ____ Make sure the oven and hood vent have been installed and properly vented. Check the vent damper; make sure it works freely without binding.

27) ____ Make sure the trash compactor (if applicable) has been installed, leveled, and squared to the opening. Make sure its legs have been properly adjusted and screwed securely to the cabinet.

28) ____ Make sure the trash compactor plug reaches an outlet.

29) ____ Check for leaks under the kitchen sink. Inspect hot and cold water lines, traps, and angle stops. Check garbage disposal and dishwasher lines. Make sure hot water tap is on the left and cold is on the right.

30) ____ Make sure the windows slide freely and latch properly.

31) ____ Make sure the screens have been installed and are free of damage and defects.

32) ____ Check floor coverings (as required):

    A) ____ Make sure carpet is complete, properly stretched, and trimmed, with no bad seams.

    B) ____ Make sure ceramic tile is complete and grouted.

    C) ____ Make sure wood parquet floors have been installed completely and protected.

    D) ____ Make sure base shoe is complete and painted as required.

33) ____ Make sure any appliances removed for flooring installation have been reinstalled and adjusted properly.

34) ____ Check luminous ceiling (if applicable). Look for full coverage, gaps, tight miters, and broken plastic panels.

35) ____ Check inside luminous ceiling; make sure all fluorescent bulbs have been installed.

36) ____ Patch any holes inside luminous ceiling light well. *Note:* This will keep bugs from the attic off the plastic panels.

37) ____ Make sure any luminous ceilings over six feet in length or width are secured by wire to prevent sagging in the center of the ceiling.

38) ____ Replace any damaged, chipped, cracked, or discolored plastic ceiling panels.

39) _____ Make sure a copper line and shut-off valve is run to the refrigerator opening for ice maker (if applicable).

40) _____ Check all kitchen doors for proper alignment and fit.

41) _____ Make sure all door hardware is complete, with proper latches.

42) _____ Make sure there is no missing or damaged baseboard and all nails have been set and filled.

43) _____ Make sure base shoe (as required) has been nailed properly, caulked, and painted.

44) _____ Check door casings and jambs for damage.

45) _____ Make sure electrical outlets and switch plate covers have been installed and leveled. Check for damage or cracks in face plates.

46) _____ Patch and texture holes in drywall.

47) _____ Check for full and complete paint coverage.

Bathrooms

1) _____ Check the general appearance of cabinets:
Good _____ Bad _____ .

2) _____ Make sure the cabinets are complete. Look for a smooth, even finish.

3) _____ Make sure toe-kicks are complete, with no gaps or voids. Check for missing scribe molding (if applicable). Check for complete paint or stain coverage.

4) _____ Make sure scribe molding at sides of cabinets has been properly nailed.

5) _____ Make sure all cabinet doors and drawers align properly.

6) _____ Replace any warped doors.

7) _____ Adjust cabinet door hinges; replace any sprung hinges.

8) _____ Make sure all drawers slide freely and do not bind. Replace any damaged rollers and drawer guides.

9) _____ Make sure drawers are flush against the cabinet face frame when closed.

10) _____ Check for damage to counter tops.

11) _____ Make sure all mirrors have been installed.

12) _____ Make sure mirror channel is the correct color (gold, silver, bronze). Trim any sharp edges.

13) _____ Make sure mirror channel has been cut properly. (FHA/VA channel if required.)

14) _____ Check for double mirror clips (per FHA/VA standards). Make sure they have been securely screwed or properly secured with mastic.

15) _____ Check mirror for defects, chips, scratches, black spots, cat-eyes, and distortions.

16) _____ Check plan and contract for correct thickness of mirror (one-quarter-inch plate; three-sixteenths-inch crystal; one-eighth-inch float).

17) _____ Make sure the medicine cabinets and mirrors line up at the top (if thirty-six-inch or thirty-eight-inch mirror).

18) _____ Check plumbing, faucets, and aerators.

19) _____ Turn on faucets and check under sinks for leaks. Check trap and hot and cold flex-stops. Make sure hot water tap is on the left and cold is on the right.

20) _____ Make sure electrical outlets and switch plate covers have been installed flush to the walls and level. Check face plates for pressure cracks at screw heads. Check G.F.I. circuits.

21) _____ Make sure towel bars have been installed per plan, (towel bars at fifty-four inches; towel rings at sixty inches) and properly secured.

22) _____ Make sure the toilet paper holder has been installed at twenty-four inches and properly secured.

23) _____ Check towel bars and paper holders for proper anchors. Make sure all bars have molly-bolts or toggles if backing was missed. *Note*: Plastic anchors tend to pull loose.

24) _____ Make sure luminous ceilings (if applicable) are complete and properly installed. Make sure all gaps have been caulked.

25) _____ Make sure fluorescent bulbs have been installed.

26) _____ Replace any damaged or cracked plastic panels. *Note*: Make sure plastic panels are not cut too short for ceiling openings.

27) _____ Make sure medicine cabinets have been installed square, with no apparent damage. Make sure screws have not been hammered—check for hammer tracks or bent screws inside medicine cabinets.

28) _____ Caulk any gaps or voids at the sides of the medicine cabinets.

29) _____ Check vinyl floor installation (if applicable). Make sure it has been caulked at door jambs, casings, and transition strips.

30) _____ Check ceramic tile installation (if applicable). Check grout for voids and cracks. Make sure tile has been caulked at the window. Caulk tile at tub and base intersection, as well as wall intersections, with silicone or approved sealant.

31) _____ Make sure base shoe has been installed (as required).

32) _____ Check carpet installation (if applicable). Make sure it has been properly stretched and trimmed, with no bad seams.

33) _____ Make sure toilet is set level, securely bolted to the floor, and properly centered. Make sure the base of the toilet is neatly and properly caulked.

34) _____ Make sure there are china caps covering all mounting bolts.

35) _____ Check toilet seat for damage.

36) _____ Make sure toilet flushes properly, without leaks or run-on.

37) _____ Make sure ceramic shower tile is complete and grouted. Make sure soap dish and grab are installed properly. Check shower neck escutcheon for caulking.

38) _____ Check shower door installation:

   A) _____ Make sure the head track is cut tight to the walls with no gap.

   B) _____ Make sure the jambs are plumb and caulked with approved sealant. Make sure jambs are caulked both inside and out; sills only on the outside. *Note*: Caulking the inside sill can cause mildew due to trapped water.

   C) _____ Make sure the bumpers have been installed.

   D) _____ Make sure the doors have been installed with the correct slide to prevent water leaks. *Note*: Inside track door should slide toward the shower head; outside track door should be opposite the shower head when closed.

39) _____ Check shower pan or tub for damage.

40) _____ Check plumbing mixer valve for leaks and shower head for blockage.

41) _____ Make sure the tub has a clear drain.

42) _____ Check tub window (if applicable) for proper slide and lock.

43) _____ Check for opaque or clear glass, as required.

44) ____ Check screen for defects, rips, tears, or dents.

45) ____ Make sure window stool and apron are free of defects.

46) ____ Make sure light fixtures have been installed and not damaged.

47) ____ Make sure doors operate and lock properly. Check door casings for damage.

48) ____ Check for squeaky floors.

49) ____ Check pocket doors (if applicable) for proper slide and lock. Check for drywall nails that might scratch the door face.

50) ____ Make sure closet doors (if applicable) have been adjusted to operate properly.

51) ____ Make sure skylights (if applicable) are clean and defect free—no scratches, paint, mastic, or other damage.

52) ____ Make sure heat registers are installed level, and finished.

53) ____ Make sure all baseboard has been installed and painted. *Note*: Make sure baseboard is installed behind the toilet prior to setting bowl and tank.

54) ____ Make sure mirrored doors (if applicable) have been installed free of defects and properly adjusted.

55) ____ Patch and texture holes in drywall.

56) ____ Check for full and complete paint coverage.

Bedrooms

1) ____ Make sure floors do not squeak. Walk and mark floors for repairs. Repair all floors prior to installation of floor covering.

2) ____ Make sure all doors fit, slide, lock properly, and have complete hardware.

3) ____ Check door casings and jambs for damage.

4) ____ Make sure baseboard is complete and nailed properly. Replace any damaged baseboard. *Note*: Make sure baseboard is complete inside closets.

5) ____ Check for an even reveal around all windows.

6) ____ Make sure window stools and aprons have not been damaged. Make sure all nails have been set and filled.

7) ____ Check base line for bows in walls.

8) ____ Check customer upgrade list for closet door specifications. Make sure doors have been installed and properly adjusted.

9) ____ Make sure top tracks of closets have been cut tight and square to opening (no gaps or voids over one-quarter-inch).

10) ____ Make sure top track is securely screwed to header.

11) ____ Make sure bottom track is secured. If carpeted, make sure plastic closet guides have not been removed by carpet layers.

12) ____ Check for rubber door bumpers on closet doors.

13) ____ Make sure mirrored doors (if applicable) are free of defects, rattles, chips, black spots, and scratches.

14) ____ Check french doors (if applicable) and thresholds for complete and proper installation. Make sure dead bolts on fixed doors line up with holes drilled into the threshold and door frame head.

15) ____ Make sure doors lock and latch properly, are free of defects and damage, and have complete hardware.

16) ____ Make sure sliding patio doors (if applicable) slide and lock freely. Check door sills for damage. Check below sills for missing trim or exterior wall covering.

17) ____ Make sure sliding door screens (if applicable) lock properly and are free of defects.

18) ____ Check for any extra television, telephone, or electrical outlets according to homeowner upgrade list.

19) ____ Check drywall for:

    A) ____ Nail pops

    B) ____ Bad tape joints

    C) ____ Damaged corner bead

    D) ____ Straight lines at off-angles

    E) ____ Poor or light acoustic coverage

    F) ____ Scrapes in ceiling and walls

    G) ____ Fractures, tears, or hammer holes

    H) ____ Blade marks and rough or uneven texture pattern

    I) ____ Properly sanded walls and cut-in ceiling

    J) ____ Full and complete paint coverage

Utility Room

1) ____ Make sure door is properly adjusted.

2) ____ Make sure sill is caulked or weatherstripped, if necessary.

3) ____ Make sure platform is painted and has baseboard. Check for complete caulk and paint.

4) ____ Remove trash or debris from under platform.

5) ____ Check for six-foot clearance on all sides of heating unit.

6) ____ Make sure return air grill has been caulked and painted.

7) ____ Make sure all heat ducts have been taped at joints.

8) ____ Make sure condensate line is hooked up properly.

9) ____ Patch and texture any holes in drywall.

10) ____ Make sure low and high combustion air is vented properly.

11) ____ Check for full and complete paint coverage.

First Floor Hallway

1) ____ Check baseboard and base line for straight alignment, and bows in the walls. Replace any damaged or missing baseboard. Check door casings for damage.

2) ____ Check drywall for:

    A) ____ Nail pops

    B) ____ Bad tape joints

    C) ____ Damaged corner bead

    D) ____ Straight lines at off-angles

    E) ____ Poor or light acoustic coverage

    F) ____ Scrapes in ceiling and walls

    G) ____ Fractures, tears, or hammer holes

H) _____ Blade marks and rough or uneven texture pattern

I) _____ Properly sanded walls and cut-in ceiling

J) _____ Full and complete paint coverage

3) _____ Make sure all doors line up properly.

4) _____ Check for bowed doors.

5) _____ Check for sprung or damaged hinges.

6) _____ Make sure return air grill is level and painted.

7) _____ Make sure the smoke detector is complete.

8) _____ Make sure the light fixtures are complete.

9) _____ Make sure all electrical outlets and switch plate covers have been installed flush to walls and level, with no pressure cracks at screw heads.

10) _____ Check floor coverings (as required):

A) _____ Make sure carpet is complete, properly stretched, and trimmed, with no bad seams.

B) _____ Make sure ceramic tile is complete and grouted.

C) _____ Make sure wood parquet floors have been installed completely and protected.

D) _____ Make sure base shoe is complete and painted as required.

11) _____ Make sure thermostat has been installed damage-free.

Laundry—Service Room—Washer/Dryer Area

1) _____ Make sure door closer (if necessary) has been installed and adjusted.

2) _____ Make sure there is a minimum sixty-inch clearance for the washer and dryer.

3) _____ Make sure dryer vent is free of debris, and damper works properly.

4) _____ Check the washer drain pipe.

5) _____ Make sure the hot and cold valves are level. Make sure the recessed plastic box (optional) is set square and level with chrome hose-bib handles.

6) _____ Check baseboard and base line for bows in the walls.

7) _____ Check door casings for damage.

8) _____ Make sure door flex-stops have been installed.

9) _____ Make sure doors fit and latch properly. Look for complete hardware.

10) _____ Make sure floor coverings and base shoe have been installed, as required.

11) _____ Make sure light fixture has been installed without damage.

12) _____ Check water heater platform for minimum eighteen-inch high drywall and seal. Make sure the water heater and the T/P valve work properly and drain outside the house.

13) _____ If the water heater is in the laundry room, make sure door is properly vented (six-by-fourteen-inch high and low screen vents).

14) _____ Make sure drywall margin around window has an even reveal.

15) _____ Make sure window stool and apron are not damaged.

16) _____ Make sure windows operate freely and lock properly.

17) _____ Make sure window screens have been installed free of rips, dents, and defects.

18) _____ Check cabinets for smooth, even finish.

19) _____ Make sure cabinet doors align horizontally and vertically, and have even margins.

20) _____ Make sure shelves inside cabinet are level and not damaged.

21) _____ Make sure scribe molding has been installed and properly nailed.

22) _____ Make sure counter top is securely installed and level.

23) _____ Make sure electrical outlets and switch plate covers have been installed flush to walls and level, with no pressure cracks at screw heads.

24) _____ Check drywall for:

    A) _____ Nail pops

    B) _____ Bad tape joints

    C) _____ Damaged corner bead

    D) _____ Straight lines at off-angles

    E) _____ Poor or light acoustic coverage

    F) _____ Scrapes in ceiling and walls

    G) _____ Fractures, tears, or hammer holes

    H) _____ Blade marks and rough or uneven texture pattern

    I) _____ Properly sanded walls and cut-in ceiling

    J) _____ Full and complete paint coverage

Stairwell

1) _____ Prior to floor covering, check stairs and landings for squeaks. Replace any damaged plywood at landings and stair treads.

2) _____ Make sure risers do not vary more than one-quarter inch.

3) _____ Check for correct handrail height.

4) _____ Make sure wrought iron or wood handrailing is securely installed and properly finished. Make sure lagbolts hit the handrail backing and are firmly in place.

5) _____ Make sure stair skirt (as required) has been installed and painted.

6) _____ Check stair handrail and cap for correct finish.

7) _____ Make sure the light fixture has been installed damage-free.

8) _____ Make sure skylights (if applicable) have been installed and cleaned.

9) _____ Check skylight light well for proper finish.

10) _____ Check for scratches, cracks, and other defects on skylights.

11) _____ Check drywall texture and paint in light well.

12) _____ Make sure baseboard is complete and painted. Replace any missing or damaged baseboard.

13) _____ Check drywall for:

    A) _____ Nail pops

    B) _____ Bad tape joints

    C) _____ Damaged corner bead

    D) _____ Straight lines at off-angles

    E) _____ Poor or light acoustic coverage

    F) _____ Scrapes in ceiling and walls

    G) _____ Fractures, tears, or hammer holes

    H) _____ Blade marks and rough or uneven texture pattern

I) ____ Properly sanded walls and cut-in ceiling

J) ____ Full and complete paint coverage

14) ____ Make sure carpet (if applicable) is complete, stretched properly, and trimmed, with no bad seams or loose bullnoses at stair treads.

Pre-Homeowner Walk-Through

After quality control, the builder or superintendent should make a final check of the following items. This should take place the day of the customer walk-through, two to four hours before the orientation tour.

1) ____ Make sure the entire house is clean. Remove any trash or debris. Check inside closets, cabinets, and behind doors.

2) ____ Make sure paint is complete, inside and outside. Make sure all touch-up painting is complete.

3) ____ Make sure all floors are covered, all floor coverings are complete, and all homeowner upgrades and options are installed.

4) ____ Make sure the entire house is complete and clean. Make sure the house *looks* finished.

5) ____ Make sure all windows have been cleaned, carpets have been vacuumed, and lemon oil has been applied to all stained cabinets and wood handrails.

6) ____ Make sure all mail boxes are installed, level, and painted or stained. Make sure the street address is clearly marked.

7) ____ Remove any trash or construction debris from the yard or garage.

8) ____ Make sure all walks have been swept and hosed down.

9) ____ Make sure the garage and driveway have been swept and hosed down.

10) ____ Remove any loose lumber or construction debris at garage rafters.

11) ____ Make sure all yards have been raked.

12) ____ Check all area drains and make sure they are not clogged or covered.

13) ____ Make sure address numbers are properly secured and none are missing.

14) ____ Make sure there is no water, mud, or silt build-up in the curb or gutter at the driveway approach.

15) ____ Sweep, hose, rake, and clean *everything*. Deliver the house to the homebuyer the way you would want a new home delivered to you.

Congratulations, the house is now complete!

Your dedication, hard work, attention to detail, and commitment to quality made it happen.

Typical Construction Schedule

	Days Required	Scheduled Days Used
A. Concrete schedule	23	23
B. Rough frame and mechanical schedule	40	63
C. Insulation and drywall schedule	22	85
D. Finish schedule	30	115
E. Flooring schedule	±	10
*Total on-site construction schedule		± 125 days

*Noted items are concurrent with other procedures.

Start Date	A. Concrete Schedule Slab-On-Grade	Days Required	Scheduled Days Used
	Reestablish blue tops and stake buildings	1	1
	Lay out and trench	1	2
	Clean footings; spread forms	1	3
	Set forms	1	4
	Clean out plumbing trenches	1	5
	Lay soil pipe	1	6
	Inspect soil pipe	1	7
	Backfill soil pipe	1	8
	Install utility sweeps and sleeves	*	*
	Clean footing trenches	1	9
	Install reinforcing steel and electrical ground rod	1	10
	Inspect footings and steel	1	11
	Pour footings	1	12
	Backfill and compact interior trenches	1	13
	Fine grade slab area	1	14
	Roll copper	1	15
	Lay membrane and sand	1	16
	Lay mesh	1	17
	Inspect pre-slab and copper	1	18
	Pour slabs	1	19
	Strip forms	*	*
	Sack and fill rock pockets	*	*
	Fine grade garages; form raised entries	1	20
	Pour garages and raised entries	1	21
	Strip forms	*	*
	Sack and fill as required	*	*
	Cleanup work areas; remove trash and form materials	1	22
	Fine grade—spin-off lots; pull excess to streets	1	23
	Total:	±	23 days

*Noted items are concurrent with other procedures.

Start Date	B. Rough Schedule	Days Required	Scheduled Days Used
	Windows: order window and door frames; set delivery schedule	*	*
	Drop lumber	1	1
	Lay out first floor	1	2
	Plate first floor	1	3
	Cut headers and detail	1	4
	Frame and raise first floor walls	1	5
	Plumb and line	1	6
	Set beams on the lower level	1	7
	Run joist	1	8
	Plumber: vent first floor	*	*
	run gas and water on first floor	*	*
	run waste	*	*
	Lay subfloor; sheet second floor	1	9
	Nail subfloor	1	10
	Build stairs	1	11
	Order Reminder:		
	Windows: schedule glass		
	Drywall: schedule board		
	Fiberglass tubs: schedule deliveries		
	Skylights: schedule deliveries		
	Metal fireplaces: schedule deliveries		
	Trusses: confirm deliveries		
	Siding: confirm deliveries		
	Roofing: schedule materials		
	Finish: schedule materials		
	Lay out second floor	1	12
	Plate second floor	1	13
	Cut headers; detail second floor	*	*
	Frame and raise second floor walls	1	14
	Plumb and line second floor walls	1	15

*Noted items are concurrent with other procedures.

Start Date	B. Rough Schedule	Days Required	Scheduled Days Used
	Set beams on second floor	1	16
	Trusses: field measure models	*	*
	Hang shear panels	1	17
	Plumber: vent second floor	*	*
	run waste to second floor	1	18
	run gas and water to second floor	1	19
	Cut rafters (conventional)	1	20
	Load rafters (conventional)	1	21
	Roll trusses (truss system)	1	22
	Build chimney chases	1	23
	Run fascia and starter board	1	24
	Load roof sheathing	*	*
	Lay roof sheathing	1	25
	Nail roof sheathing; nail shear panels	1	26
	Drywall: confirm delivery dates	*	*
	Secure hold-downs and anchor bolts	*	*
	Nail-off all straps	*	*
	Remove braces; scrap-out buildings	*	*
*	Inspect shear panels and hold-downs	1	27
	Build decks	1	28
	Cut out heat vents	*	*
	Heating and air-conditioning: start rough heat	*	*
	Build soffits, platforms, and dropped ceilings	*	*
	Roofer: schedule and confirm materials	*	*
	Fireplaces: install metal boxes and flues	*	*
	Cleanup: pick up scrap lumber	*	*
*	Inspect roof sheathing	*	*
	Concrete: chip for screed and drypack garage jambs	*	*
	Roofer: run felt	1	29
	Sheet metal: run screed, vents, and roof metal	1	30

*Noted items are concurrent with other procedures.

Start Date	B. Rough Schedule	Days Required	Scheduled Days Used
	Deliver and install fiberglass bathtubs	*	*
	Plumber: complete rough and prep for top-out inspection	1	31
	Inspect rough plumbing	*	
	Electric: start rough (layout, box, and rope)	1	32
	Drywall: schedule and confirm delivery dates	*	*
	Roofer: load roofs	*	*
	Windows: install windows and door frames	1	33
	Framer: complete exterior elevations	1	34
	Finish carpenter or framer: set wood door frames	*	*
	Finish carpenter: order and schedule hardware	*	*
	Insulation: insulate behind tubs and showers	1	35
	Telephone: prewire buildings	1	36
	Lath: start wrapping buildings	1	37
	Cleanup: sweep all buildings; draft-stop and rat-proof	1	38
	Inspections:	2	40
*	rough plumbing, top-out		
*	rough heating and air-conditioning		
*	rough electric		
*	framing		
	Order Reminder:		
	Cabinets: measure and schedule		
	Light fixtures: order and schedule		
	Total	±	40 days

*Noted items are concurrent with other procedures.

Start Date	C. Insulation and Drywall Schedule	Days Required	Scheduled Days Used
	Drywall: stock board ± (ten lots)	1	1
	Insulation: start lids; start walls	1	2
	Inspect insulation	1	3
	Cleanup: dig around buildings prior to plaster	*	*
	Windows: install glass and sliding doors (three-trip)	*	*
	Appliances: schedule appliances	*	*
	Drywall: start lids and clips	1	4
	start walls	2	6
	run metal; complete all nailing	1	7
	scrap-out buildings	1	8
	Inspect drywall and lath nailing	1	9
	Plaster: apply scratch coat (minimum five lots)	1	10
	Garage doors: measure and schedule	*	*
	Plaster: apply brown coat	1	11
	Paint: spray overhangs	*	*
	Cleanup exteriors after brown coat	*	*
	Drywall: start tape	1	12
	top and fill	1	13
	box and spot nails	1	14
	apply finish coat	1	15
	final point-up and drying time	1	16
	sand and prepare for texture; mask beams	1	17
	texture	1	18
	cleanup (scrape floors, clean windows, and hose out buildings)	1	19
	scrape and blade walls	1	20
	Painter: seal kitchens and baths	1	21
	enamel kitchens and baths	1	22
	Total:	±	22 days

*Noted items are concurrent with other procedures.

Start Date	D. Finish Schedule	Days Required	Scheduled Days Used
	Cabinets: deliver a minimum of five to seven lots	1	1
	start installations	1	2
	Windows: order and schedule screens	*	*
	Weatherstrip: order and schedule	*	*
	Fiberglass tubs: schedule repairs (minimum three-week notice)	*	*
	Plaster: start color coat	1	*
	Fencing: order materials; schedule start date	*	*
	Finish carpenter: deliver materials	1	*
	spot materials; start shelf and pole	1	3
	hang interior doors	1	4
	hang exterior doors	1	5
	run base	1	6
	Wrought iron: measure handrails	*	*
	Painter: seal exterior doors (as soon as possible after hanging)	*	*
	Roofer: lay roof material (as soon as possible after exterior complete)	*	*
	Marble tops: measure pullman tops	*	*
	Luminous ceilings: measure; schedule deliveries	*	*
	Shower doors: measure pullman tops	*	*
	Mason: start fireplace faces	*	*
	Cabinets: pick up rough-top prior to ceramic tile	*	*
	Ceramic tile: walk jobsite; inspect and accept prior to start	*	*
	run paper, mesh, and metal	1	7
	float kitchen counters	1	8
	lay tile	1	9
	grout tile	1	10
	Plaster: remove staging	*	*
	Garage doors: install doors	*	*
	Wrought iron: confirm deliveries	*	*

*Noted items are concurrent with other procedures.

Start Date	D. Finish Schedule	Days Required	Scheduled Days Used
	Fireplaces: order glass doors	*	*
	Concrete: grade for flatwork and sidewalks	1	11
	Concrete: lay base and compact city walks	1	12
	Concrete: form walks and rear stoops	1	13
	Concrete: pour walks	1	14
	Drywall: pickup prepaint	1	15
	Mail boxes: order mail boxes	*	*
	Cleanup: sweep for paint	*	*
	Painter: putty and caulk woodwork	1	16
	undercoat woodwork	1	17
	paint walls	1	18
	complete exteriors	1	19
	enamel interior trim and woodwork	1	20
	Fiberglass tubs: schedule second (final repairs, tub buffs)	*	*
	Grader: start fine grade; pull excess to the street	1	21
	Wrought iron: install handrails	1	22
	Finish carpenter: install handrail cap	1	23
	Painter: prep handrails and entry doors	*	*
	stain handrail and lacquer entry doors	*	*
	Heating and air-conditioning: start finish; set furnaces	*	*
	Marble tops: install bath and pullman tops	*	*
	Mirrors: measure mirrors and mirrored wardrobe doors	*	*
	Cleanup: sand floors, as required	*	*
	Finish carpenter: install weatherstrip	*	*
	run hardware	*	*
	Fireplaces: install glass doors	*	*
	Appliances: deliver appliances	*	*
	Light fixtures: deliver fixtures	*	*
	Heating and air-conditioning: set air-conditioning units	*	*
	Cleanup: clean tubs and showers	*	*

*Noted items are concurrent with other procedures.

Start Date	D. Finish Schedule	Days Required	Scheduled Days Used
	Cleanup: clean windows (as soon as possible after fine grade)	*	*
	Finish electric: install plugs and switches	1	24
	hang fixtures; set appliances	1	25
	Fencing: set fence posts	*	*
	Windows: hack-out, as required	*	*
	Fiberglass tubs: complete final repairs and tub buffs	*	*
	Water district: order water meters	*	*
	Street signs: complete signage	*	*
	Finish plumbing: start trim; set appliances	*	26
	Luminous ceilings: install kitchen and bath ceilings	*	*
	Civil engineer: certify lot and final monuments	*	*
	Fencing: complete required fencing	*	*
	Shower doors: install doors	*	*
	Mirrors: install mirrors and mirrored wardrobe doors	*	*
	Windows: install screens; adjust slide and lock	*	*
	Garage doors: install openers as required	*	*
	Inspect: final grading	*	*
	Cleanup: sweep, box-out, haul trash	*	*
	Framer: repair floors (prior to covering)	*	*
	Superintendent's final punch list: (ongoing through finish)	1	27
	Final subcontractor repairs	1	28
	Painter: voluntary paint pickup	1	29
	Inspect final building	1	30
	Utilities: follow-up on utility releases and meter hookups	*	*
	Total:	±	30 days

*Noted items are concurrent with other procedures.

Start Date	E. Flooring Schedule	Days Required	Scheduled Days Used
**	Loan approvals: confirmed by sales	*	*
	Order and delivery of flooring materials	3	3
	Resilient flooring: install sheet goods	1	4
	Wood flooring: install flooring	*	*
	Ceramic flooring: lay tile floors	1	5
	grout tile	1	6
	Carpet flooring: strip and pad	1	7
	lay carpet	1	8
	Cleanup: final	1	9
	Complete final quality control	*	*
	Pick up final paint	*	10
	Total	±	10 days

*Noted items are concurrent with other procedures.

Sample Pay Schedule Ratios

Appliances

Delivery	90%
Installation complete	10%

Cabinets

Delivered and installed	90%
Retention	10%

Carpentry—Finish

Exterior frames installed	10%
Interior materials delivered	10%
Interior materials installed	70%
Retention	10%

Carpentry—Rough—Labor

First floor plumb and line	30%
Second floor plumb and line	20%
Roof sheathing	20%
Frame inspection	20%
Retention	10%

Carpentry—Rough—Lumber and Trusses

First drop	65%
Final delivery	35%

Ceilings, Luminous

Installation complete	90%
Retention	10%

Ceramic Tile

Surrounds complete	45%
Counters/decks complete	45%
Retention	10%

Clean-Up

Framing clean-up	25%
Plaster clean-up/windows washed	45%
Finish clean	20%
Retention	10%

Closets

Installation complete	90%
Retention	10%

Concrete—Flatwork

Pour complete	90%
Retention	10%

Concrete—Slabs

Trench/formed	40%
Footing poured	40%
Slab/garage poured	10%
Retention	10%

Curb and Gutter

Field measure	90%
Retention	10%

Drywall

Board stocked	35%
Nailing inspection	35%
Texture/clean-up	20%
Retention	10%

Electric

Sleeves/stubs installed	10%
Rough inspection	50%
Finish complete	30%
Retention	10%

Fireplace—Boxes

Rough installation/cap set	80%
Doors set/adjusted	10%
Retention	10%

Fireplace—Faces

Material delivered	30%
Veneer/face complete	60%
Retention	10%

Flooring

| Installation complete | 90% |
| Retention | 10% |

Garage Doors

| Installation complete | 90% |
| Retention | 10% |

Grading—Rough

| Field measured | 90% |
| Retention | 10% |

Hardware

| Installation complete | 90% |
| Retention | 10% |

Heating and Air-Conditioning

Rough inspection	60%
Equipment and finish set	30%
Retention	10%

Insulation

| Installed/certification complete | 90% |
| Retention | 10% |

Landscape

| Complete | 90% |
| Retention | 10% |

Light Fixtures

| Delivery | 90% |
| Retention | 10% |

Mirrors

| Installation complete | 90% |
| Retention | 10% |

Painting

Exterior prime	20%
Interior prime	20%
Interior/exterior complete	50%
Retention	10%

Paving

| Field measure | 90% |
| Retention | 10% |

Plumbing

Ground work complete	40%
Top-out complete	30%
Finish	20%
Retention	10%

Roofing

Paper/load	60%
Lay field	20%
Mud ridges and hips	10%
Retention	10%

Sheet Metal

| Installation complete | 90% |
| Retention | 10% |

Shower-Tub Enclosures

| Delivery and installation complete | 90% |
| Retention | 10% |

Storm Drain

| Field measure | 90% |
| Retention | 10% |

Structural Steel

| Installation complete | 90% |
| Retention | 10% |

Stucco

Lath inspection	40%
Brown coat	40%
Color coat	10%
Retention	10%

Water Systems

| Field measure | 90% |
| Retention | 10% |

Weatherstripping

| Installation complete | 90% |
| Retention | 10% |

Windows

Deliver windows—install door frames/panels	80%
Install/adjust screens	10%
Retention	10%

Glossary

Bogies: Uneven texture or finish; clumps and humps that need to be bladed, scraped, or sanded flat.

Brown coat: Second coat of plaster or stucco operation; applied over scratch coat.

Color: Third and final coat of plaster or stucco; applied over brown coat.

Cross-sight: Line up with an adjacent surface for level and plumb.

Crown: The high point of a span, such as a joist, rafter, or beam.

Dig back: Exposing foundation at ground level for application of waterproofing, plaster, or finish.

Dog-legged: Bowed or uneven; not straight.

Dual glazing: Double pane thickness of glass to increase insulation factor of a window; also called insulated glass.

FHA/VA: Federal Housing Authority/Veterans Administration standards or guidelines.

Firewall: Wall constructed with a fire resistance rating; five-eighths-inch, type "X" drywall over two-by-four studs in a one hour rated firewall.

G.F.I.: Ground fault interrupt circuit; used to prevent electrical shock at damp or wet locations, such as baths and kitchens.

Green board: Drywall that is moisture resistant, used in wet locations around tubs and showers.

I.C.B.O.: International Committee of Building Officials.

Isometric: Plan or layout detailing pipe and duct dimensions and length of runs, such as plumbing and gas systems.

Notice to proceed: Written authorization to subtrades to start work; should establish start date and production rate.

P.S.I.: Pounds per square inch; used for measuring leaks in water or gas pipes.

P.V.C.: Poly vinyl chloride; plastic pipe used for water lines, conduits, or sleeves.

Plant-ons: Architectural details, such as exposed trim or masonry, to add definition and character to elevations.

"R" value: Rating given to construction materials for their insulation factor and resistance to heat loss.

Schematic: Wiring or mechanical layout of circuits or heat system.

Scratch coat: First coat of plaster or stucco operation; applied over lath.

Screed: Edge or ground to which concrete, plaster, or stucco is leveled.

Setback: Distance between foundation and established point, usually to property line.

Shear nailing: Additional nailing required on plywood or drywall to increase structural strength and stability of the building.

Shinners: Nails or fasteners that protrude through finished materials; need to be chipped flush, counter sunk, puttied, and repainted.

Slump test: Test for moisture in concrete in which concrete is piled in a mound (nine to twelve inches high) to see if it retains its shape.

T/P valve: Test and pressure valve; valve off of hot water heater to release pressure buildup and test line.

U.B.C.: Uniform Building Code; recognized standards to be followed in house construction.

110-V: Standard voltage circuit for standard household lighting, plugs, and switches.

220-V/240-V: Special voltage circuit for major equipment and appliances, such as air-conditioners and ovens.